Lubrication and Lubricant Selection

Lubrication and Lubricant Selection

A Practical Guide

by

A R Lansdown

Mechanical Engineering Publications
London and Bury St Edmunds

First published 1996

© A R Lansdown

ISBN 1 86058 029 7

A CIP catalogue record for this book is available from the British Library.

Printed in Great Britain by Antony Rowe Ltd, Chippenham, Wiltshire

Preface

In the thirteen years which have elapsed since the first edition of this book was published (Lansdown, 1982, *Lubrication: a practical guide to lubricant selection*, Pergamon Press), there have been few major changes in lubricant technology. There has, however, been considerable steady progress in the temperature capabilities of lubricants, and in the service life expected of many lubricants. This is noticeable in the case of vehicle engine oils, where the operating temperatures have increased by 15–20 °C, while at the same time oil change periods have doubled to 15 000 kilometres or more.

Perhaps the greatest changes have been those brought about by greater public concern for safety and for the care of the environment. This has resulted in new legislation governing such subjects as product liability and environmental protection. Lubricant technologists have played, and are continuing to play, an important part in reducing energy consumption and in reducing pollution, both by gaseous emissions and by used lubricants.

This edition has been completely updated to cover all these developments. The lists of test methods have been revised to take into account the greater degree of international standardization, as well as the amendment and improvement of published methods which is continually taking place.

Greater coverage has been given to the nature and selection of lubricating oils, which represent about ninety percent of all lubricant consumption. An emphasis has also been put on automotive oils, which are a very high proportion of the oil used.

Finally, the chapter on lubricant monitoring has been completely re-written to accommodate the wider interest in the subject and the much greater variety of monitoring methods available.

Related Titles

High temperature lubrication ISBN 0 85298 897 4
A R Lansdown

Lubrication of gearing ISBN 0 85298 831 1
W J Bartz

Lubricants in operation ISBN 0 85298 831 1
U J Moller and U Boor

A Tribology casebook ISBN 1 86058 041 6
J D Summers-Smith

Journal of engineering tribology ISSN 1350/6501
Proceedings of the Institution of
Mechanical Engineers – Part J

For the full range of titles published by MEP contact:

Sales Department
Mechanical Engineering Publications Limited
Northgate Avenue
Bury St Edmunds
Suffolk
IP32 6BW
England

Tel: 01284 763277
Fax: 01284 704006

About the Author

Dr A. R. Lansdown trained originally as a chemist, and received his first degree at University College, Cardiff, and his doctorate at Cambridge. From Cambridge he went to Canada as a National Research Council Post-Doctorate Fellow.

In 1956 he joined the Production Research and Technical Service Laboratory of Imperial Oil Limited in Calgary. Here he became involved for the first time in the application of chemistry to various aspects of engineering, especially the flow of fluids through the rock formations of petroleum reservoirs.

In 1961 he returned to Britain to join the Ministry of Aviation, with responsibilities for aircraft lubricants as well as hydraulic fluids and fuels. Like many other lubricant technologists at that time, he soon realised that the purely chemical aspects of lubricant formulation and behaviour could not be divorced from the engineering aspects of lubrication, the realization which was soon afterwards to lead to the concept of tribology as a unifying discipline.

His period in the Ministry of Aviation coincided with the intensive search for high-temperature lubricants and other products for use in supersonic aircraft, and he became particularly interested in synthetic and solid lubricants.

In 1968 he was appointed Manager, and subsequently Director, of the newly-formed Swansea Tribology Centre, and retained this post until his retirement in 1988. In addition to the management of the Centre, he also carried out a wide variety of consultancy projects, including a number of plant failure investigations, and these further involved him in analysing the engineering aspects of the failures.

He took an active part in the activities of the Institute of Petroleum and the Tribology Group of the Institution of Mechanical Engineers. He was a member of the Council of the Institute of Petroleum from 1984 to 1990. He was awarded the Tribology Silver Medal in 1986 and became a Fellow of the Institution of Mechanical Engineers in 1993.

Dr Lansdown has widely published in the fields of lubrication and tribology, his most recent book High Temperature Lubrication (Mechanical Engineering Publications Limited, 1994) proving to be highly successful.

Since his retirement from the Swansea Tribology Centre he has continued to work as an independent consultant.

Contents

Chapter 1

Basic Principles of Lubrication

1.1 Meaning of lubrication

This book is intended as a practical guide to lubricants and lubrication, and hence it deals almost entirely with practical subjects. The proper selection and application of lubricants will, however, be made easier if a little of the basic theory is understood. This first chapter therefore tries to explain the basic theory in a brief and simple manner.

When one surface moves over another, there is always some resistance to movement; the force which opposes movement is called friction. If the friction is low and steady, there will be smooth, easy sliding. At the other extreme the friction may be so great, or so uneven, that movement becomes impossible or the surfaces can overheat or be seriously damaged.

Lubrication is simply the use of a material to improve the smoothness of movement of one surface over another; the material which is used in this way is called a lubricant. Lubricants are usually liquids or semi-liquids, but may be solids or gases or any combination of solids, liquids, and gases.

Generally speaking, the smoothness of movement is improved by reducing friction. This is not, however, always the case, and there may be situations in which it is more important to maintain steady friction than to obtain the lowest possible friction. Some examples are: the control of chatter in a machine tool slideway or grinding operation; control of strip movement in metal rolling; and elimination of brake squeal or clutch judder in a car.

In addition to reducing or controlling friction, lubricants are usually expected to reduce wear and often to prevent overheating and corrosion. These secondary requirements will be considered later.

1.2 Friction

In most cases dry friction between two bodies closely follows two laws which are usually called Amontons' Laws, although in fact they were first described by Leonardo da Vinci in about 1500.

The first of these laws states that the friction between two solids is independent of the area of contact. If, for example, a brick slides over a flat surface, then according to this law the friction will be the same whether it is sliding on its base, its side, or its end.

The second law is that the friction is proportional to the load exerted by one surface on the other. To continue the example of the brick, if a second brick is placed on top of the first, the friction will be doubled; with a third brick the friction will be trebled, and so on.

It is this second law which makes it possible to define the coefficient of friction. If the friction is proportional to the load then, in other words, the friction is equal to a constant multiplied by the load

$$F = \text{constant} \times W$$

or

$$\frac{F}{W} = \text{constant}$$

This constant is the coefficient of friction, usually written μ. The coefficient of friction mainly depends on the materials which are in sliding contact, and can vary from 0.03 to 3.0 or more.

The coefficient of friction between two bodies is, in fact, not quite constant. It often varies with change in load, and usually varies with sliding speed. The static friction – the force needed to start one body sliding over another – is almost always greater than the dynamic friction – the force needed to keep it moving at the same speed once sliding has started.

The friction between two dry surfaces arises from two main basic sources: adhesion and deformation. The most important of these is adhesion.

It is difficult to imagine that, for example, two smooth pieces of cast iron can stick together in the same way as adhesive tape, but in fact all materials will stick together to some extent. This may be easier to imagine if we remember that even the smoothest engineering surfaces are rough when viewed under a high-powered microscope. This roughness means that two solid surfaces which are touching one another are only in contact at the peaks of their asperities. This is shown in an exaggerated form in Fig. 1.1.

The asperities are very small, something like a millionth of a square millimetre or a thousand-millionth of a square inch in area. If two

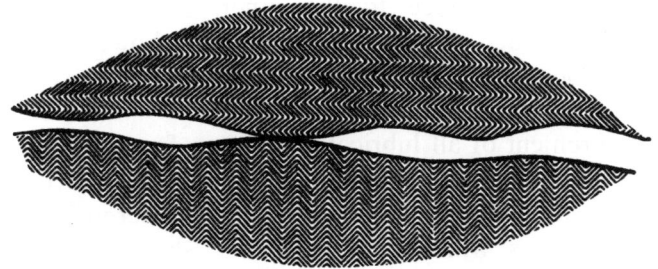

Figure 1.1   Contact of two solid surfaces

surfaces touch, even with a load of only a few grams, the load will be carried entirely on a small number of these asperities, and the actual contact pressure on the asperities may be as high as $1400 \, \text{MN/m}^2$ (200 000 psi). Such enormous pressures can squeeze the asperities on hardened steel or cast iron surfaces out of shape, and even make them weld together.

Under such pressures it is easy to imagine asperities sticking together, even if they do not actually weld. Once they have stuck together, force is needed to separate them. The lateral force required determines how great the adhesive friction will be.

The other main component of friction – deformation friction – is usually small, and it is obvious why this must be so. The deformation which takes place when two surfaces rub together must be either elastic (temporary) or plastic (permanent). If the deformation is elastic, then the energy which is used to produce deformation will be recovered when the surface resumes its original shape, and there will be no net friction. If the deformation is plastic, then there will be a permanent change in shape, and the amount of change which can be tolerated in any engineering component will be limited.

There is, in fact, one further complication in that when some substances deform elastically, there is a delay, or 'hysteresis', in their return to their original shape. When this happens, the energy released after contact may be too late to be returned to the sliding system, and there will then be a net frictional loss. This can be important for some plastics and especially some rubbers. For example, 'high hysteresis' rubbers for vehicle tyres are deliberately formulated to produce high hysteresis friction. This is because tyres lose most of their adhesive friction when wet, and high hysteresis ensures that the deformation friction is sufficient to provide good grip on wet roads.

For accuracy it should perhaps be noted that when two asperities adhere they will often separate by a rupture, or tearing, inside one of the surfaces. This results in a very small amount of material being transferred from one surface to the other, so that there is some deformation taking place. The primary influence, however, is adhesion, and if this is reduced or eliminated, the resulting deformation will also be reduced or eliminated.

The first requirement of all lubrication is, therefore, the elimination or reduction of the force required to shear the adhesive junctions formed between asperities. This can be done either by interposing a material between the asperities which is more easily sheared, or by using a chemical which will alter the shear stress of the asperities themselves.

The material interposed between asperities may be a gas, a liquid, or a solid. If it is a gas or a liquid, or a semi-liquid such as a grease, then a third type of friction is introduced, this is **viscous friction** – the force necessary to shear a viscous fluid.

1.3 Liquid lubrication

The way in which liquids lubricate can be simply explained by considering the example of a plain journal bearing, as shown in Fig. 1.2. As the shaft (journal) rotates in the bearing, lubricating oil is dragged into the loaded zone and the pressure and volume of the oil in the

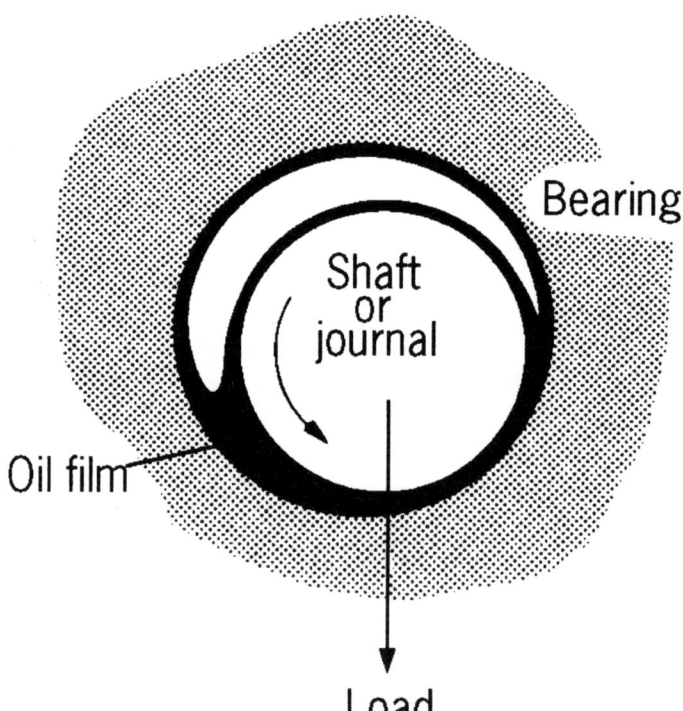

Figure 1.2   Lubrication of a plain journal bearing

loaded zone both increase. The pressure rise, and the thickness of the oil film, depend on the shaft speed and the lubricant viscosity. The relationship between speed, viscosity, load, oil film thickness, and friction can be understood by considering a graph such as the one in Fig. 1.3.

In this graph, the coefficient of friction is plotted against the expression ZN/P where

$$\frac{ZN}{P} = \frac{\text{oil viscosity} \times \text{shaft speed}}{\text{bearing pressure}}$$

There are three distinct zones in the graph, separated by the points A and B. At B the coefficient of friction is at its minimum, and this is the point at which the oil film is just thick enough to ensure that there is no contact between asperities on the shaft and bearing surfaces. In Zone 3, to the right of B, the oil film thickness is increasing, because of increasing viscosity, increasing speed, or decreasing load. The coefficient of friction increases as the oil film thickness increases. Zone 3 is the zone of 'hydrodynamic' lubrication.

As conditions change from B towards A, the oil film thickness decreases so that the asperities on shaft and bearing rub against each other. The amount of rubbing, and the friction, increase as the oil film thickness decreases.

At A the oil film thickness is reduced virtually to nil, and the load between shaft and bearing is being carried entirely on asperity contacts. In Zone 1, to the left of A, the coefficient of friction is almost independent of load, viscosity, and shaft speed. Zone 1 is the zone of 'boundary' lubrication.

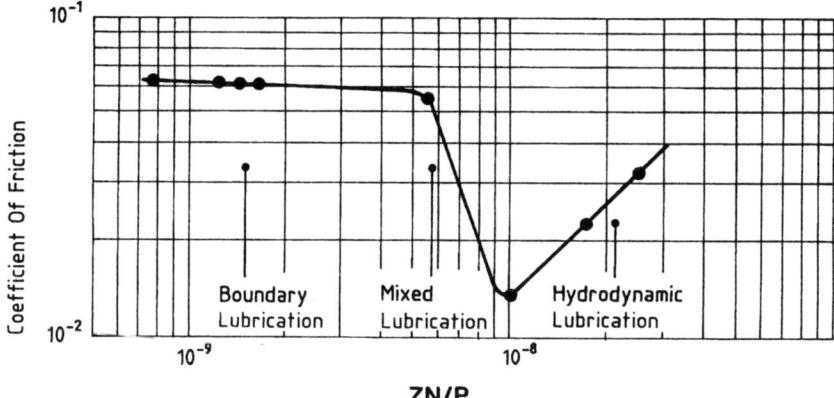

Figure 1.3   Stribeck curve

Zone 2, between A and B, is known as the zone of mixed lubrication. The shaft load was traditionally considered to be supported by a mixture of oil pressure and asperity contacts, so that the lubrication was a mixture of 'hydrodynamic' and 'boundary' lubrication. The term 'mixed lubrication' may, in fact, be even more appropriate than was originally thought, because modern theories suggest that in different systems there may be a mixture of four or more different types of lubrication present in this zone.

The different lubrication zones also have an important influence on wear. Wear is a very complicated subject, with several different forms, more than one of which may be present at the same time. This will be discussed briefly in section 1.8. Generally speaking the amount of wear which takes place depends on the severity with which two surfaces rub against one another.

In Zone 3 there is no contact between the surfaces and, therefore, there is no wear. As the oil film becomes thinner in moving through Zones 2 and 1, there is increasingly severe contact between the surfaces and, therefore, a greater tendency to wear.

As far as lubricant selection is concerned, the two important zones in Fig. 1.3 are Zones 1 and 3, the zones of boundary lubrication and hydrodynamic lubrication. This is because finding the best solutions for Zones 1 and 3 will also provide the best solution for Zone 2.

1.4 Hydrodynamic lubrication

Hydrodynamic lubrication simply means lubrication which is achieved by movement of a liquid. In the case of the plain journal bearing which has been used as an illustration, the rotation of the shaft causes lubricant to move into the loaded zone. Since the loaded zone will be the point at which the shaft and bearing surfaces are closest together, the entry into this zone is tapered, like a curved wedge.

As the oil is forced to move into the narrower part of the wedge, the pressure increases, and it is this 'hydrodynamic pressure' which supports the shaft load. Increasing load reduces the oil film thickness, while increasing hydrodynamic pressure increases the oil film thickness. The hydrodynamic pressure in turn is determined by the viscosity of the oil and the speed at which it is squeezed into the wedge-shaped zone.

A similar pressure wedge is necessary in almost all systems which rely on hydrodynamic lubrication. For example, in a straight sliding bearing the wedge may be produced by a tilting of the slider, as shown in Fig. 1.4. Alternatively, a sort of 'mini-wedge' may be obtained by radiusing, chamfering, or relieving the leading edge of the slider. In some cases a wedge may be generated on a perfectly flat slider because the centre of the sliding face warms up in use and expands, thus

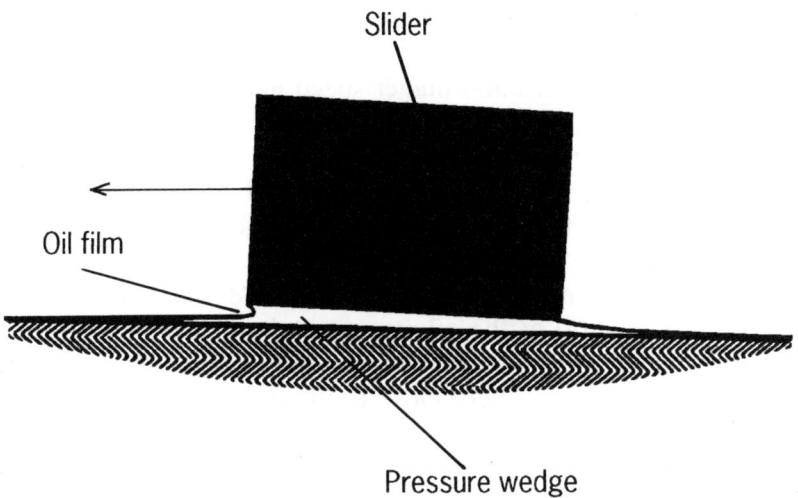

Figure 1.4 Pressure wedge in a slider bearing

producing a 'thermal crowning'. Any or all of these types of wedge may be present, for example in a pad-type thrust bearing, but some type of wedge is usually essential to produce hydrodynamic lubrication.

The one exception is where one bearing surface is moving towards the other, so that the lubricant between the two surfaces is squeezed, and forced to move out of the space between them. The viscosity of the lubricant then tends to prevent the lubricant from being squeezed out. The higher the viscosity of the lubricant, the greater is its resistance to being squeezed out of the bearing and, therefore, the greater the protection against damage to the bearing surfaces. This is known as a squeeze film effect.

The situation is described mathematically by the Reynolds equation

$$\frac{\delta}{\delta x}\left(h^3 \frac{\delta P}{\delta x}\right) + \frac{\delta}{\delta z}\left(h^3 \frac{\delta P}{\delta z}\right) = \eta\left(6U\frac{\delta h}{\delta x} + 6h\frac{\delta U}{\delta x} + 12\right)$$

where
   h is the lubricant film thickness,
   P is the pressure,
   x and z are the coordinates,
   U and V are the speeds in directions x and z.
   The terms $6U(\delta h/\delta x)$ and $6h(\delta U/\delta x)$ describe the rate at which the oil is being squeezed into the wedge
   η is the oil viscosity.

For most users, however, it is sufficient to remember that the oil film thickness depends on the speed of the bearing surfaces and on the oil viscosity. *The viscosity is the only property of the oil which is important in hydrodynamic lubrication.*

In hydrodynamic lubrication higher speed gives better lubrication, and very low speed may cause lubrication failure.

The oil film in pure hydrodynamic lubrication should ideally be just thick enough to ensure that there is no contact between the asperities on the two surfaces. In other words, the oil film thickness should be greater than the sum of the asperity heights. In Fig. 1.3 this ideal point is B, but because the speed and load of a bearing, and the temperature (and therefore the viscosity) of the oil, cannot be kept absolutely constant, it has been usual to aim at a point just to the right of B. This not only ensures that the friction is very close to the minimum possible, but also ensures that the wear is kept to a minimum.

In practice, new bearing surfaces will often be rougher than is desirable, and a small amount of contact may be permitted. This will allow wear to take place until the roughness is reduced, or the surfaces are 'run-in'. The aim, however, has usually been to design the bearing system so that once it is run-in, the oil film thickness is greater than the sum of the asperity heights.

This traditional approach has been intended to minimize or eliminate wear at the expense of a slight increase in the average friction. The increasing importance of reducing fuel consumption, especially in vehicle engines, has recently led to a change of emphasis. Now the target in vehicle engines is to achieve the lowest possible average friction, at the expense of a small amount of wear.

This result is obtained by the use of less viscous lubricating oils calculated, for example, to ensure average operation at point B in Fig. 1.3 instead of to the right of B. This provides the lowest possible viscous friction, and the resulting adhesive friction and wear are minimized by the use of suitable additives in the oil.

A special type of hydrodynamic lubrication can take place in certain heavily loaded contacts, such as in ball or roller bearings and many types of gear. If the geometry and the type of movement are suitable, the lubricant can be trapped in the entry zone to such a contact and become subjected to very high pressures as it is squeezed into the confined space in the most highly loaded part of the contact.

These pressures have two important effects. First, they cause the viscosity of the lubricant to increase considerably, thus increasing its load-carrying capacity. At the same time they cause the loaded surfaces to deform elastically in such a way as to spread the load over a greater area. Because the load-carrying capacity is controlled by elastic and hydrodynamic effects, the phenomenon is known as elasto-hydrodynamic lubrication. Figure 1.5 gives an impression of how elastohydrodynamic lubrication takes place.

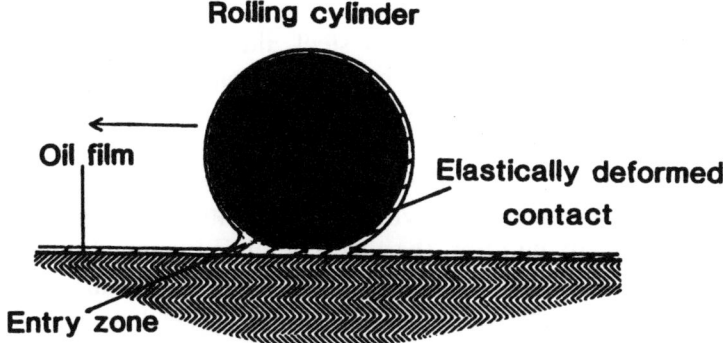

Rolling cylinder

Oil film

Elastically deformed contact

Entry zone

Figure 1.5   Elasto-hydrodynamic lubrication of a cylinder on a flat surface

Several different equations have been produced to enable the the film thickness in elasto-hydrodynamic lubrication to be calculated. One of the earliest and best known of these equations is the Dowson–Higginson Equation

$$h = \frac{2.65(\eta_0 U)^{0.7} R^{0.43} \alpha^{0.54}}{E^{0.03} P^{0.13}}$$

where:
   η_0 is the viscosity of the lubricant in the entry zone
   R is the effective radius
   E is a material parameter
   P is a contact pressure parameter
   α indicates the extent to which viscosity increases with pressure
For most users it is sufficient to remember that the important property of the lubricant is, once again, its viscosity.

Although, technically, hydrodynamic lubrication refers to a form of lubrication by liquid, it can also be applied to lubrication by gas, provided that the load and speed conditions are suitable for the very low viscosity of gases. However, elasto-hydrodynamic lubrication cannot be obtained with gases.

1.5   Boundary lubrication

Whenever the lubricant film thickness becomes too small to give full fluid film separation of the surfaces, surface asperities start to come into contact with one another, and lubricant properties other than the bulk viscosity start to become important. In Zone 1 in Fig. 1.3, the oil film has become so thin that there is no hydrodynamic contribution and only boundary lubrication is effective.

In most normal situations, asperities are initially coated with a film of oxide, such as iron oxide on iron or steel, aluminium oxide (alumina) on aluminium, and so on. When such asperities rub together, their tendency to adhere is relatively mild. However, if the oxide film is removed by severe rubbing, the exposed metal surfaces have a very powerful tendency to adhere. Thus, if the bearing surfaces retain their oxide films, the contact between asperities will give moderate friction and mild wear. If they lose their oxide films, there will be high friction and severe wear. In either case the object of boundary lubrication is to reduce friction and wear. There are several ways in which this can be done.

(a) *Adsorption*

All solid surfaces tend to attract a thin film of some substance from their environment. Such films may be only one or a few molecules thick, and are said to be 'adsorbed' onto the surface. The process is shown diagrammatically in Fig. 1.6, in which molecules of a long-chain alcohol are depicted adsorbed onto a metal surface.

The strength of adsorption depends on the electronic structure, and 'polar' molecules – those in which there is a variation in electronic charge distribution along the length of the molecule – tend to adsorb with their molecules perpendicular to the surface.

Thicker or more strongly adsorbed films give greater protection to the

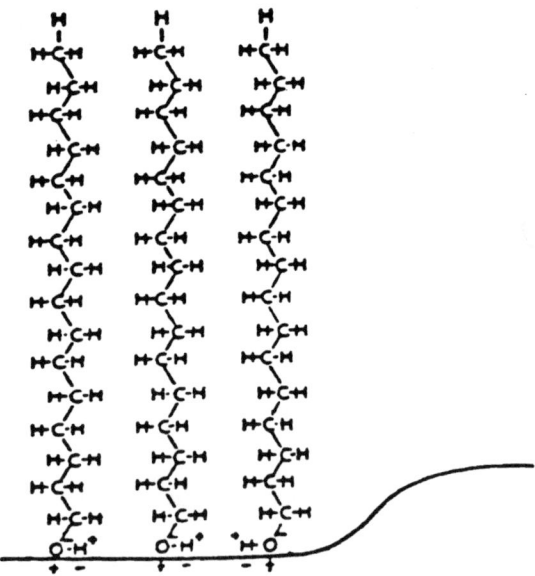

Figure 1.6   Adsorption of a long-chain alcohol

bearing surface. Thus it follows that the preferred boundary lubricants for adsorption onto a metal bearing surface are long-chain, polar organic chemicals. These will be considered in a later chapter.

Adsorption is a reversible process. An adsorbed substance can be desorbed if heated to a critical temperature, or displaced by a substance which is more strongly adsorbed. This latter effect is valuable in boundary lubrication, because the most strongly adsorbed substances present in the lubricant will be preferentially adsorbed. In a well-formulated lubricant these will be the more effective boundary lubricants.

A useful by-product of adsorption is believed to be the so-called 'Rehbinder Effect'. This is a reduction of the modulus and yield stress of metals in the presence of an adsorbed film. As a result of this effect, lower stress will be developed when asperities collide. In particular, during the running-in of new bearing surfaces, the removal of excessive asperities will take place more mildly.

(b) *Chemisorption*

After adsorption of some substances onto a metal surface, an exchange of bonding electrons can take place with the metal or oxide surface to produce a new chemical compound. Such substances are said to be 'chemisorbed'. Figure 1.7 shows the chemisorption of stearic acid on an iron-containing surface as iron stearate as an example.

Chemisorbed materials are more strongly bound to the metal surfaces than are adsorbed materials, and the chemisorption process is not easily reversible. Long-chain molecules again orient themselves perpendicular to a surface. Such films can give very effective boundary lubrication.

(c) *Chemical reaction*

Adsorbed and chemisorbed films are very effective in reducing friction and mild wear under light or moderate rubbing. They are fairly easily removed mechanically under severe rubbing conditions and are therefore not very effective in preventing severe wear or seizure. The natural oxide layer reduces severe wear and seizure, but once it has been removed by rubbing, the re-oxidation of the surface may be too slow to be effective.

To handle such situations, more reactive chemicals can be added to the lubricant to react with the bearing surfaces and produce protective films. Suitable films include: chlorides, sulphides, phosphides and phosphates; and any chemical which will react with a bearing surface to produce such substances will be effective in producing protective films. The problem is, however, that really reactive chemicals, such as

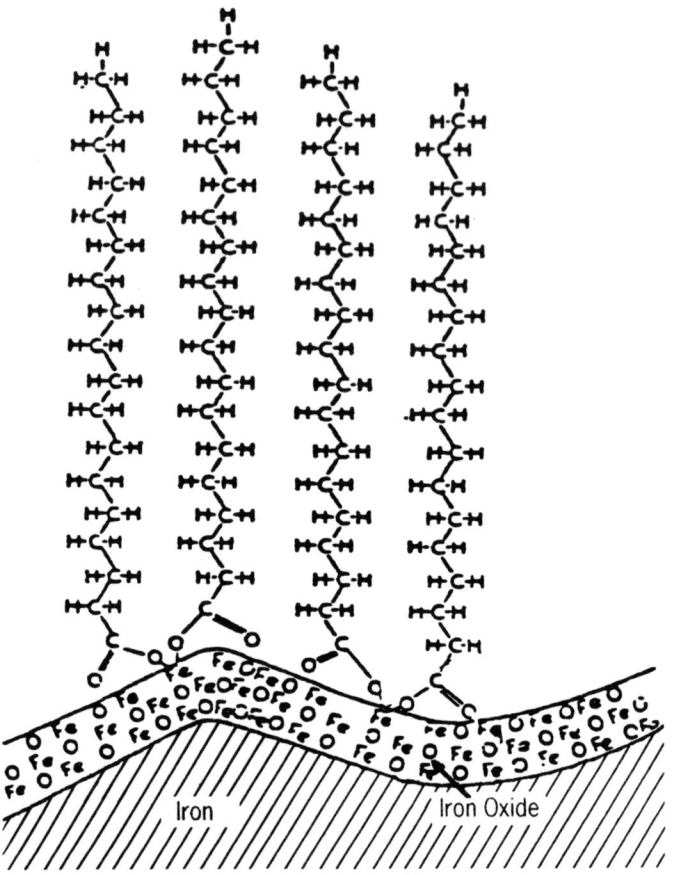

Figure 1.7 Chemisorption of stearic acid on iron

hydrochloric or phosphoric acid, will continue to react and will thus corrode away the metal surface.

The solution is to use organic compounds which contain sulphur, phosphorus, or chlorine. These may adsorb or chemisorb on oxidized metal surfaces but will react rapidly with freshly exposed metal surfaces from which the oxide film has been removed. For example, trixylyl phosphate will chemisorb onto oxidized bearing steel, but under severe rubbing will react with freshly exposed steel to produce surface layers of iron phosphate or phosphide.

In this way the reaction of the chemical with the bearing surface can be limited to the minimum necessary for lubrication, and corrosion is controlled. Even so, some of the more powerful chlorine-containing extreme pressure additives will attack some metals. These should only be used where rubbing conditions are very severe, such as in metal cutting.

An indication of the effectiveness of such films is given by the use of processes such as phosphating and sulphiding on new rubbing surfaces to prevent seizure during running in.

1.6 Externally pressurized lubrication

Section 1.4 explained how the pressure necessary to give full fluid film separation of loaded bearing surfaces is produced by the movement of the surfaces. The same effect can be obtained by forcing the lubricant into the bearing under an externally-applied pressure. This will enable full fluid film separation to be achieved where the viscosity or speed would be insufficient to support the load hydrodynamically.

The basic theory of externally-pressurized (sometimes called hydrostatic) lubrication is very simple. The average pressure required is equal to the load divided by the effective bearing area

$$P = \frac{W}{A}$$

In practice, the design of hydrostatic bearings must also take into account the need to maintain stability and to control lubricant flow.

External pressurization may be used with a liquid lubricant or even a grease. It is also commonly used with a gas, where it can offset the problems associated with the very low viscosity of gases.

1.7 Dry or solid lubrication

A solid lubricant is basically any solid material which can be placed between two bearing surfaces and which will shear more easily under a given load than the bearing materials themselves. The coefficient of friction in dry lubrication is related to the shearing force and the bearing load

$$\mu = \frac{\text{Shear force}}{\text{Load}}$$

In practice, certain other properties are needed for a good solid lubricant, such as chemical stability, ability to adhere to a bearing surface, and so on. This is described more fully in Chapter 7.

1.8 Wear

There are several different types of wear, all of which have different causes and effects. The subject is, however, too complicated to be explained in detail in this book. In practice, the most common forms of wear are either adhesive or abrasive. It may be helpful to explain briefly the nature of adhesive wear and abrasive wear, and how they are affected by lubricants.

Adhesive wear is caused by exactly the same type of adhesion as described in Section 1.2. The severity of adhesive wear depends on the size of the adhesive forces. The resulting damage may consist of the transfer of only a few molecules, transfer of a whole asperity, fatigue fracture of an asperity, breaking away of a fragment due to overstressing, or melting of the surface.

A lubricant which reduces the adhesive forces between two surfaces will obviously tend to reduce adhesive wear. Referring again to Fig. 1.3, the lubricant will completely eliminate wear in Zone 3, and reduce it in Zone 2.

The situation in Zone 1 is not quite so clear-cut. Lubrication will normally reduce wear in this zone, and may even virtually eliminate it in special cases. Where the contact load or speed or both are very high, a severe form of adhesive wear called **scuffing** may take place. The resulting damage may be worse than it would be without any lubricant. The use of various additives can be particularly effective in Zone 1. This subject is explained in Chapter 3.

Abrasive wear, or simply abrasion, is the wear which is caused by a hard rough surface cutting or ploughing material from a softer surface. The rough surface may be continuous as with the surface of a grindstone, in which case the wear is called two-body abrasion. Alternatively it may consist of separate particles such as sand or carborundum powder, in which case the wear is called three-body abrasion.

In either case the use of a lubricant is more likely to increase wear than to reduce it. There are two situations in which a lubricant can help. The first is where the lubricant is able either to prevent abrasive particles from entering the contact zone or even to flush particles out of the system. The second is where the lubricant film is thick enough for the abrasive surface or abrasive particles to pass through the loaded zone without contacting the other surface. In other situations the lubricant may increase wear by holding the abrasive particles together in the form of a grinding paste. This can actually increase the amount of wear debris produced by each abrasive contact.

It follows that if wear is taking place in a machine it is important to try to establish what sort of wear it is before making any decision about lubrication. One way of identifying the type of wear is to examine the wear debris, and this is discussed in Chapter 11.

1.9 Cooling

The cooling effect of a lubricant arises in two ways. Reducing friction will reduce the amount of heat generated in a bearing. In this way any material, liquid, solid or gas, which reduces friction will also act as a coolant.

More often, however, the cooling effect is produced by removing excess heat from one part of the system and transferring it to some other part where it can be lost to the outside air or to a separate cooling system. For this purpose, the important factors are a high specific heat of the lubricant and a high flow rate. Liquids are better coolants than gases because of their higher specific heats. Greases and solids are quite ineffective because of lack of flow.

1.10 Corrosion prevention

It is probably because mineral oils are very efficient in preventing corrosion that there is a tendency to expect all lubricants to do the same. Most liquid lubricants and greases will give some corrosion protection, but some of the synthetic lubricants and vegetable oils are less efficient in this respect than mineral oils.

Lubricants which contain water, such as soluble cutting oils and fire-resistant emulsions, may be actively corrosive, and solid lubricants can vary from mildly protective to mildly corrosive. Lubricants which are initially non-corrosive, or even protective, can also become contaminated with water or acids in service so that they become corrosive.

When wear and corrosion take place together, the combination is called corrosive wear. The resulting loss of material can be much greater than when both effects occur separately. This is because: (a) a corroded surface may be much less wear-resistant than an uncorroded surface; and (b) a worn surface may be much more easily corroded than an unworn surface. Corrosive wear is usually more easily cured by preventing corrosion than by trying to prevent the wear.

It is important to assess the corrosion risk in any system, and to ensure that, where necessary, lubricants are formulated with corrosion inhibitors. This subject is discussed in more detail later.

1.11 Summary

To sum up, lubricants are needed to give smooth sliding and to prevent over-heating and damage to moving surfaces. Lubricants may be liquids, solids, or gases.

The most important property of a liquid lubricant is its viscosity. However, where contact between the moving surfaces cannot be completely prevented, the boundary lubricating properties will also be important.

Lubricants can often also assist in cooling and in preventing corrosion of the lubricated parts.

Chapter 2

Choice of Lubricant Type

2.1 The problem of lubricant selection

If there were only a few lubricants available, the problem of lubricant selection would hardly arise. The available oil, grease, or solid lubricant would be used, and one would put up with its disadvantages and whatever life it gave. This was generally the situation before the mid-nineteenth century, when people used such things as mutton-fat, goose grease, crude oil from seeps, natural graphite, or even molybdenite (crude natural molybdenum disulphide) – whichever happened to be locally available.

Nowadays, the variety of lubricants is enormous. Most lubricant manufacturers can supply scores of different mineral oils, many different greases, and several different synthetic oils, and there are scores of manufacturers in Britain alone, as well as specialized suppliers of anti-seizes and dry lubricants.

This great variety means that many of the simpler machines can be lubricated satisfactorily by any one of hundreds of different lubricants. For example, a door hinge or latch may work properly with a wide range of mineral oils, synthetic oils, greases, dry lubricants, or, for quite a while, with none at all. Even much more complicated machines may not be very critical in their lubricant needs. It may be because of this that machines are often completely designed and built before any thought is given to their lubrication.

The choice of lubricant is, in fact, often a sub-conscious one, following previous practice or what is felt to be normal practice for the particular type of machine or component. For example, it would be almost automatic to use oil in a machine tool gearbox or grease in an isolated ball-bearing. In most cases no problem arises, and the machine operates satisfactorily with the lubricant chosen as an afterthought.

However, there have also been instances where none of the hundreds of available lubricants were suitable. When this happens, there may be an urgent and expensive rush to develop a suitable lubricant or to re-design the machine. It is not unknown for a large or expensive machine to be scrapped because it could not be lubricated.

More often it is found that while the lubrication of the various components is acceptable, there are operating problems relating to the choice of lubricant, such as high lubricant consumption, leakage, short lubricant life, or excessive fuel consumption.

The problem of lubricant selection should therefore be considered early in the design of a machine. It is, however, a fact of life that machines often reach their users with lubrication problems still unsolved. Other factors may lead to a reassessment of the lubricant requirements of a machine. New and better lubricants will often become available, or the original lubricant become unavailable. A machine may be modified, or used for a different purpose, or moved into a different environment. The selection of lubricants is therefore an important subject for users of machines as well as their manufacturers to understand.

In spite of the above, there are good technical and commercial reasons for using in a new machine the lubricants which have been specified by the manufacturer. It is normally only when a problem arises, or a change in operation is considered, that the lubricant selection needs to be re-assessed. The one important exception to this is where rationalization of the lubricant supply becomes desirable. This is considered in Chapter 4.

In choosing a lubricant for a particular application, the object should be to obtain the lowest overall long-term cost; this definitely does not mean using the cheapest available lubricant. It is no good using a cheap oil if, as a result, the oil or the machine breaks down, or the maintenance staff have to spend too much time checking and replacing the oil. In other words, reliability is likely to be far more important than the initial price of the lubricant.

2.2 Basic types of lubricant

Lubricants are usually divided into four basic classes.

(a) *Oils*   A general term used to cover all liquid lubricants, whether they are mineral oils, natural oils, synthetics, emulsions, or even process fluids.

(b) *Greases*   Technically these are oils which contain a thickening agent to make them semi-solid. It is convenient, however, to include the anti-seize pastes and the semi-fluid greases under the same heading.

(c) *Dry lubricants* These include any lubricants which are used in solid form, and may be bulky solids, paint-like coatings, or loose powders.

(d) *Gases* The gas usually used in gas bearings is air, but any gas can be used which will not attack the bearings, or itself decompose.

The broad properties of the four classes of lubricants are summarized in Table 2.1. Their detailed properties are described in later chapters.

The advantages and disadvantages of oils stem from their ability to flow easily. Thus, on the credit side, it is very easy to pour them from a container, to feed them into a bearing by dripping, splashing or pumping, and to drain them out of a machine when no longer fit for use. Most important of all, with proper design the bearing itself can be made to feed oil into the loaded zone. Other advantages are the cooling of a bearing by carrying away heat, and cleaning it by removing debris.

On the debit side, oil can equally easily run away from a bearing, leak out of a container or machine, migrate away over the surrounding surfaces, and even evaporate if the bearing is hot and well ventilated. In addition, because it is liquid, it does not form an effective seal against dirt or moisture getting into the bearing.

The behaviour of greases is very similar to that of oils, but the former are used where the advantages of easy flow are outweighed by the disadvantages. Thus greases do not easily leak out of a machine or container, do not migrate away, and will form an effective seal against contaminants. On the other hand, greases are much less easy to feed into a bearing and are almost totally useless for cooling.

The advantages and disadvantages of solid lubricants are rather like the extremes for greases, where the lubricant will not flow at all. Similarly, the advantages and disadvantages of gas lubricants are like the extremes of oils, where the flow properties are almost too good. Solid lubricants and gas lubrication have very specialized advantages and disadvantages, and are generally used only in special situations.

2.3 Choosing the lubricant class

A lubrication system should never be made more complex than it needs to be in order to work properly. The simplest possible technique will often be the most reliable, as well as the cheapest. One of the best ways to select a lubricant – and a lubrication system – will, therefore, be to start with the simplest possible arrangement and only to alter it where this is necessary to overcome problems.

The simplest form of lubrication consists of a small quantity of plain mineral oil in place in a bearing, with no oil feed system. This will satisfy the needs of a vast range of lubricated systems, from door hinges

Table 2.1 Properties of basic lubricant types

Lubricant property	Oil	Grease	Dry lubricant	Gas
1. Hydrodynamic lubrication	Excellent	Fair	Nil	Good
2. Boundary lubrication	Poor to excellent	Good to excellent	Good to excellent	Usually poor
3. Cooling	Very good	Poor	Nil	Fair
4. Low friction	Fair to good	Fair	Poor to good	Excellent
5. Ease of feed to bearing	Good	Fair	Poor	Good
6. Ability to remain in bearing	Poor	Good	Very good	Very poor
7. Ability to seal out contaminant	Poor	Very good	Fair to good	Very poor
8. Protection against atmospheric corrosion	Fair to excellent	Good to excellent	Poor to fair	Poor to good
9. Temperature range	Fair to excellent	Good	Good to excellent	Excellent
10. Volatility	Very high to low	Generally low	Low	Very high
11. Flammability	Very high to very low	Generally low	Generally low	Depends on gas
12. Compatibility	Very bad to good	Fair to good	Excellent	Generally good
13. Cost of lubricant	Low to very high	Fairly high to very high	Fairly high	Generally very low
14. Complexity of bearing design	Fairly low	Fairly low	Low to high	Very high
15. Life determined by	Deterioration and contamination	Deterioration	Wear	Ability to maintain gas supply

Table 2.2 Possible lubricant choices when a small quantity of plain mineral oil ceases to cope with the bearing requirements

Problem	Possible solution
1. Load too high	– More viscous oil
	– Extreme pressure oil
	– Grease
	– Dry lubricant
2. Speed too high (which may make temperature too high)	– More oil or oil circulation
	– Less viscous oil
	– Gas lubrication
3. Temperature too high	– Additive or synthetic oil
	– More viscous oil
	– More oil, or oil circulation
	– Dry lubricant
4. Temperature too low	– Less viscous oil
	– Synthetic oil
	– Dry lubricant
	– Gas lubrication
5. Too much wear debris	– More oil, or oil circulation
6. Contamination	– Oil circulation system
	– Grease
	– Dry lubricant
7. Longer life needed	– More viscous oil
	– Additive or synthetic oil
	– More oil, or oil circulation
	– Grease

and locks to sewing machines and bicycles. It will fail when the speed, load, or temperature is too high, the life needed is too long, or the oil becomes too dirty with wear debris or other contaminants. When this situation arises, the problem of lubricant selection starts. It may then be necessary to use a different oil, a grease, a dry lubricant, a gas lubricant, or a lubricant feed system. Table 2.2 shows some of the possible choices when the use of a small quantity of plain mineral oil is no longer good enough.

The two main factors affecting the choice of lubricant class are usually the speed and the load. Figure 2.1 shows the speed and load limits for different types of lubricant. The boundaries in this are rather arbitrary and are, therefore, drawn very thick. The actual limit of load and speed for a particular application will depend on the type of component involved, as well as the nature of the particular lubricant. For example, the maximum specific load for a grease at low speeds can vary from $2\,000$ kN/m^2 (about 280 psi) for a plain soft grease to $6\,000$ kN/m^2 (about 850 psi) for an EP grease or molybdenum disulphide grease.

Similarly, the upper speed limit for dry lubricants is shown as about 500 mm/s (about 100 ft/min) because they are generally poor conductors of heat and tend to overheat at higher speeds. Where a dry lubricant, such as lead, has better thermal conductivity, it may be usable at higher speeds.

The figure should therefore be considered as a broad guide to lubricant selection, and not as an exact indicator.

Gas lubrication has not been shown in Fig. 2.1 for simplicity, and its scope is described in Chapter 8. However, as a general rule gas lubrication tends to be used at high speeds and low loads.

There is, in fact, a general tendency to move to lower viscosity at high speed and to higher viscosity at high load, as shown diagrammatically in Fig. 2.2. It must be remembered, however, that this is a very much over-simplified picture, and should only be considered as a very general guide.

Other factors may in certain cases completely outweigh those mentioned above. For example, in high vacuum a solid lubricant may be the only one which will not evaporate, while in textile manufacture a white solid lubricant or a gas might have to be used to avoid all risk of contamination of the product.

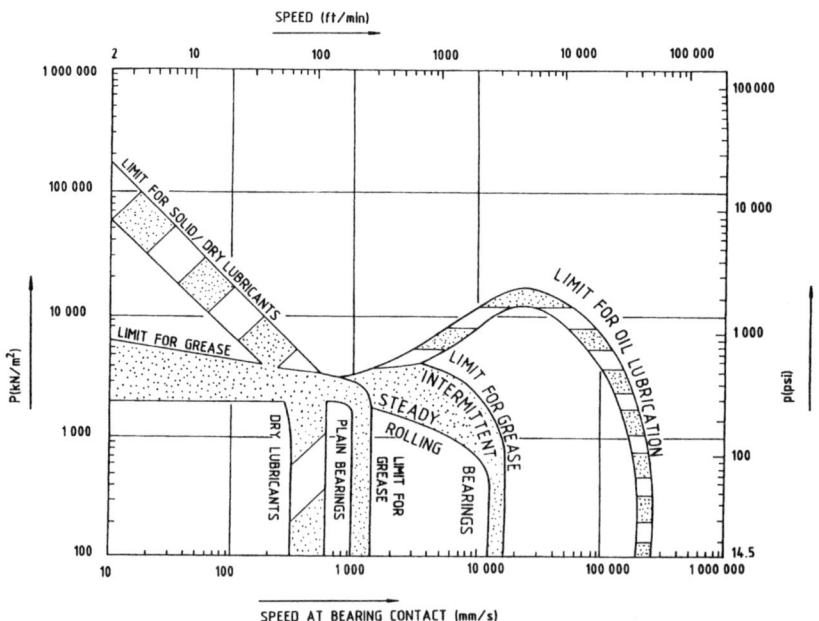

Figure 2.1 Speed and load limits for different lubricant classes

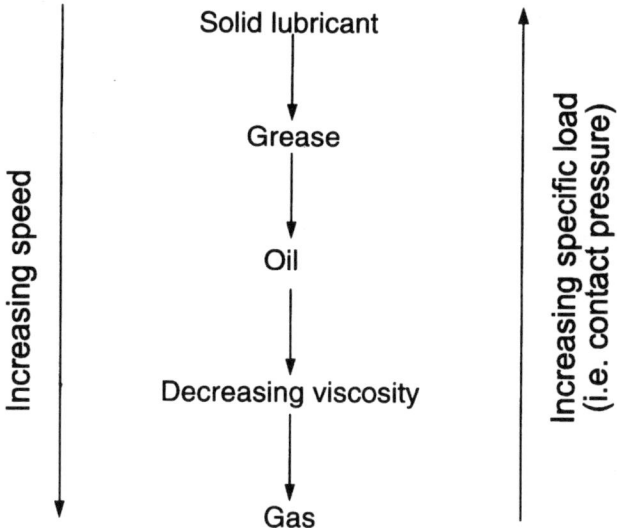

Figure 2.2   Effect of speed and load on lubricant choice

2.4   Lubricant choice for particular components

The choice of lubricant class for a number of specific components will now be discussed to show how the various requirements are considered in relation to the properties of the lubricants. A general summary of the extent to which different components need different lubricant properties is given in Table 2.3.

2.4.1   *Plain bearings*

Ideally, plain bearings are lubricated by a fluid lubricant, where full fluid film lubrication gives complete separation of the bearing surfaces, so that friction is low and no wear takes place. If the speed is high, oil is preferred, because it will give lower friction and can carry away the frictional heat. At lower speeds grease may be acceptable, because there is no heating problem and the grease will be easier to retain in the bearing. If there is a problem of contamination by dust or dirt, grease will be better at sealing out the contaminants. If the speed is so high that oil has to be used, then it may be necessary to use a full oil circulation system and filter the oil to remove the dirt.

At extremely high speeds in small bearings, gas lubrication may be preferred, but the construction requires very precise machining, and the load will be limited.

If the temperature is too high or too low for oil or grease, dry lubrication may be the best solution. Dry lubricants, however, always

Table 2.3 Requirements of different components

Type of bearing	Plain journal	Rolling bearing	Closed gears	Open gears, ropes	Clock and instrument
Lubricant property					
1. Boundary lubricating properties	*	*	***	***	**
2. Cooling	**	**	***		
3. Low friction	*	*	**	**	**
4. Ability to remain in bearing	*	**		***	***
5. Ability to seal out contaminants		**		**	
6. Temperature range	**	**	**	*	
7. Protection against corrosion	**	**	*	***	

wear away in use so that their life is limited, and their friction may be higher than with oil or grease. Solid lubricants are often preferred for very high loads and very low speed, such as in bridge bearings.

2.4.2 Rolling contact (ball or roller) bearings

Probably far more rolling bearings are lubricated by grease than by oil. This is because grease has the advantage of providing a very effective seal in the bearing covers. This prevents loss of lubricant or entry of contaminants, while at the same time maintaining an adequate supply of lubricant to the moving surfaces. Once again, oil must be used at the higher speeds, but grease can be used to much higher speeds in rolling bearings than in plain bearings, as shown in Fig. 2.1.

Gas lubrication cannot be used for rolling bearings, but dry lubrication can. Some of the techniques used are described in Chapter 8.

2.4.3 Enclosed gears

Enclosed gears are usually oil lubricated. For most applications plain mineral oils are satisfactory, but at higher speeds lower viscosities and anti-oxidant additives should be used. Hypoid gears and heavily loaded spiral bevel gears will need EP additives, while worm gears benefit from low friction additives. At low speeds greases are sometimes used, especially in small gears, and thixotropic oils or semi-fluid greases have been increasingly used in recent years.

As with rolling bearings, gas lubrication cannot be used in gears. Chapter 8 describes ways in which dry lubrication can be used.

Enclosed gears are always supported by bearings, and the same lubricant must usually satisfy the needs of the bearings and the gears.

2.4.4 Steam turbines

Steam turbines are oil lubricated, and have very critical requirements for oil properties. The temperatures are high, speeds are high, and water is present. Reliability is particularly important because a steam turbine may be very large, is often a major central feature of complex plant, and is expected to operate for long periods – often years – without maintenance or oil change.

Due to the high temperature and speeds there is a strong tendency for oxidation, so that highly refined oils with anti-oxidants are used.

Condensation of the steam leads to water contamination, and because of the high temperature and speed, this tends to form an emulsion. Centrifugal separators are used to remove the water but the oil must be formulated for easy separation, and must be kept clean by filtration and generally clean housekeeping.

There is also a tendency for rusting, because water is present, and rust inhibitors are added to protect the metal surfaces in the system.

High speed also tends to cause foaming, which can be harmful in two ways.

(a) If foam enters a bearing it will be a less effective lubricant, and bearing damage may result.
(b) Because foam occupies a larger volume than the oil which is in it, any excessive foaming will lead to loss of oil from the system.

Steam turbine oils are therefore formulated to reduce the amount of foaming.

2.4.5 Open gears, ropes, chains

Although oils are usually used on open gears, wire ropes, and chains it is important to make sure that the oil is not thrown off with high speed or sudden acceleration. The oils used are therefore viscous and 'tacky', and are bituminous or contain 'tackiness' additives. Greases will often give acceptable lubrication, but tend to be too expensive for large-scale use in these applications.

2.4.6 *Clock and instrument pivots*

Traditionally clock and instrument pivots have been lubricated by a small quantity of oil in the bearings. Originally this was vegetable oil, replaced during the nineteenth century by mineral oil. Synthetic oils are now preferred, formulated to give low viscosity with low volatility and some boundary lubrication capability. The loads and speeds are so low that, according to Fig. 2.1, any of the lubricant types should be suitable. For these applications, however, very low friction is needed, and a low-viscosity oil is most suitable.

For some instruments in which low friction is particularly important, such as gyroscopes in inertial navigation systems, gas bearings are now being used.

2.4.7 *Hinges, locks, latches, etc.*

These devices are just a few common examples of the vast numbers of small, simple moving components in which sliding or rubbing takes place. They are usually low speed, but some, such as spring-loaded latches or switches, firing pins in guns, or circuit breakers, may move very quickly for a brief time.

The contact pressures may be surprisingly high, especially in those devices containing some type of spring/trigger arrangement, such as circuit breakers and spring latches. It may be vital, therefore, to ensure that some lubricant is present, otherwise switches may break, door-locks and bolts stick, and guns jam.

Apart from the general rule that light oils are preferred where rapid motion takes place, the choice of lubricant may not be critical, and the so-called 'general purpose' oils are widely used for these unspecialized uses. In fact, the choice of lubricant is often decided for some reason not directly concerned with the quality of lubrication, such as:

(a) the interior mechanisms of door-locks and car window winders will usually be grease-lubricated in manufacture because they are so inaccessible that they are hardly ever re-lubricated, and grease will last longer than oil;

(b) door-hinges and latches of refrigerators and washing machines use dry lubricants to avoid contamination of food or clothes;

(c) switches and circuit-breakers either use dry lubricants or are coated with metals which will not stick, because oils or greases can cause interference with the electrical contacts.

2.5 Lubrication of complex machines

A complex machine may include a large number of different lubricated components. An internal combustion vehicle engine contains pistons, cylinders, plain bearings, gears, cams, cam followers, tappets, valves, and belt or chain drives. A steel rolling mill contains motors, gears, universal couplings, plain bearings, rolling bearings, pistons, cylinders, and the contacts of the rolls themselves.

In selecting the class of lubricant for such complex machinery, a further important consideration is the extent to which it is necessary, desirable, or possible to use a common lubricant for several or all of the components present. In the case of the rolling mill, the decision is likely to be to use at least three – and probably more – different lubricants. In the case of the internal combustion engine the decision will almost always be to use a single oil for all the components. However, if we consider the whole vehicle as a single machine, there will probably be at least four different lubricants in use.

There are a wide range of choices available to the designer, but the absolute limit on choice is that the lubricant must be suitable for each component in which it is used.

This chapter has indicated the factors which need to be considered in deciding whether a component should be lubricated with an oil, a grease, or even a solid or gas. The following chapters will show in greater detail how to choose the individual type of grease, oil, solid or gas, and how to use them.

Chapter 3

Lubricating Oils

3.1 The important properties of oils

The most important single property of a lubricating oil is its viscosity. The reason for this is explained in Chapter 1. As far as the actual lubrication is concerned, the only other important factor is the boundary lubrication quality. Many other properties are important for other reasons, but if at any instant the combination of viscosity and boundary lubrication is satisfactory, then the oil will perform properly as a lubricant.

In practice, many other factors must be taken into account, to make sure that the oil continues to lubricate properly over the required life, or to make sure that nothing else goes wrong with the system, and so on. Some of the other factors are listed below.

(a) *Thermal – or temperature – stability*
  If an oil becomes hot in use, then it is important that the heat does not make it break down so much that it ceases to lubricate properly.
(b) *Chemical stability*
  An oil can be chemically attacked by oxygen from the air, by water, or by any other substances with which it comes into contact. Such attack can also make it unsuitable for use. Chemical stability means the ability to resist chemical attack, and must be assessed in relation to the substances which the oil is expected to contact. Chemical stability is related to temperature because the speed of a chemical reaction increases as the temperature increases.
(c) *Compatibility*
  This is a more general term referring to any interaction between the oil and other materials present. For example, an oil may cause the rubber of a seal to swell, shrink, soften, or harden.

(d) *Corrosiveness (corrosivity)*
     Corrosion is a particular type of incompatibility in which the oil, or something in the oil, attacks a metal. An oil which is completely non-corrosive when new may become corrosive after a period of use.

(e) *Thermal or heat conductivity*
     This is important where the oil is required to conduct heat away from the bearing.

(f) *Heat capacity (specific heat)*
     This is important because it determines the amount of heat which the oil can absorb from the bearing or carry away by convection or circulation.

(g) *Flammability*
     It is obviously important that the oil should not catch fire under the conditions under which it is used. This is particularly important in some industries such as aviation and coal-mining. (Flammability means the same as inflammability.)

(h) *Toxicity*
     This is a very broad term, meaning almost every respect in which a substance can affect health. It is discussed in Chapter 14.

(i) *Availability*
     It is obvious that an oil must be available before it can be used. However, where a piece of equipment is to be used in different parts of the world, availability may become both important and difficult.

(j) *Price*
     It is also obvious that price is an important aspect of lubricating oil, or of almost anything else, but it is not a simple aspect. A single bearing failure in a critical position may cost more than a company's whole lubricant bill for a year. So, while price competition cannot be ignored, the choice of the right type of oil should never be ignored for price reasons alone. Only where different oils meet the same specification can a choice be based on price alone. Even then, it is important to make sure that the specification is fully relevant to the application for which the oil is being selected.

All these factors will be considered in more detail later in this and subsequent chapters.

Mineral oils are easily the most widely used lubricating oils. Consequently, they are often the standard with which other oils are compared. They will therefore be used to illustrate much of the rest of this chapter. It may be helpful to consider first what mineral oils are, and some of the terms used in describing them.

3.2 Mineral oils

The word 'mineral' originally meant relating to mines, and if an oil well is thought of as a sort of small-bore mine, then the name is very appropriate. Mineral oils generally mean oils obtained from petroleum, although they can also be obtained from similar sources, such as oil shales and tar-sands.

The mineral oils used for lubrication were originally just the fractions, obtained by distilling petroleum, which had a suitable viscosity for lubrication. They were, in fact, much the same as heavy fuel oils. During the last sixty years their manufacture has become much more sophisticated, with vacuum distillation, 'sweetening' to remove sulphur compounds, hydrogenating to produce more stable oils, dewaxing, isomerization, and so on.

The chemical compounds which comprise mineral oils are mainly hydrocarbons, which contain only carbon and hydrogen. These are of three basic types. The majority in any lubricating oil are paraffins, as shown in Figs 3.1(a) and (b), in which the carbon atoms are in straight or branched chains – but not rings. The second most common type are naphthenes, in which some of the carbon atoms form rings, as shown in Fig. 3.1(c). Finally, there is usually a small proportion, perhaps two percent, of aromatics, in which carbon rings are again present, but the proportion of hydrogen is reduced, as shown in Fig. 3.1(d). Both the number of carbon atoms in a ring and the alternate single and double bonds give special properties to aromatic compounds.

Apart from these four types of hydrocarbon, there may be small quantities of compounds present which also contain other elements, such as oxygen, sulphur, phosphorus and nitrogen. These compounds are sometimes referred to as asphaltenes.

For the user there are a few points which are worth remembering about these various types of compound.

(a) If the amount of carbon present in paraffin chains is much higher than the amount in napthene rings, the oil is called a 'paraffinic' oil.

(b) If the proportion in naphthene rings is not much less than the proportion in paraffin chains, the oil is called 'naphthenic'.

(c) Although the amounts of aromatics and asphaltenes present are always small, they can play an important part in boundary lubrication and oxidation stability.

The compositions of a typical paraffinic oil and a typical napthenic oil are shown in Table 3.1.

Different degrees of refining result in different proportions of the various components. The aim of the most severe modern refining

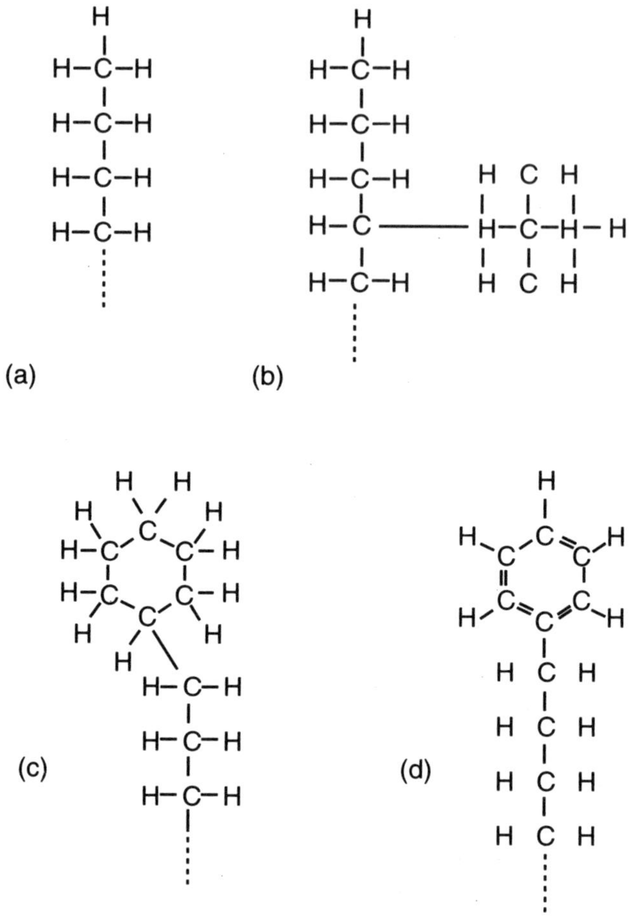

Figure 3.1 Main types of hydrocarbon

Table 3.1 Compositions of typical paraffinic and naphthenic oils

Constituents	Percentages present	
	Paraffinic oil	*Naphthenic oil*
Carbon atoms in paraffin chains	63	52
Carbon atoms in naphthene rings	33	44
Carbon atoms in aromatic rings	2	2
Sulphur (by weight)	0.5	1
Asphaltenes	1	2

processes is to provide lubricant base oils with a high proportion of branched-chain paraffins. Cracking processes reduce the average molecular weight of the hydrocarbons present and open up some or all of the naphthene rings. Reforming alters the proportion of branched-chain paraffins. Hydrogenation eliminates 'unsaturated' double bonds and aromatic groups.

Together these processes can produce base oils which consist almost entirely of branched-chain paraffins. These may be known as 'highly-refined' or 'severely refined' oils. They are still mineral oils, but in some respects their properties are more like those of synthetic hydrocarbons.

Depending on the source of the crude oil and on the severity of the refining processes, mineral base oils are now available which range in character from the traditional paraffinic or naphthenic oils to the most severely-refined types.

3.3 Viscosity

The viscosity of a liquid is its resistance to flow. Under the same conditions, a liquid with low viscosity flows more quickly than a liquid of high viscosity. Water is a liquid of low viscosity, while syrup and molasses are liquids of high viscosity.

Viscosity is measured in several different units; the relationship between the common units is explained in the Appendix at the end of this chapter. For practical purposes, attempts are being made to concentrate on the unit of 'kinematic viscosity' – the centistoke (cSt) (millimetre^2 per second, mm^2 s^{-1}). In this book viscosities are generally given in centistokes. Water at 20 °C has a viscosity of approximately 1 cSt, while a 15W/40 motor oil may have a viscosity of about 200 cSt when it is poured out of a can at 20 °C.

The viscosity of any liquid decreases as the temperature increases. The rate at which it changes varies with the type of oil, and Fig. 3.2 shows the variation of viscosity with temperature for several different mineral and synthetic oils.

Obviously, in selecting the best oil viscosity for any application, it is the viscosity at the usual operating temperature which is most important. However, if the equipment will often have to make a cold start, it is also important that the viscosity at starting temperature is not at a level which would impede starting of the machine. It follows that it is important to know how much the viscosity will change with temperature. The best way to show this is by a graph, such as in Fig. 3.2. A much more widely used method, however, is by means of its 'Viscosity Index,' or VI.

The viscosity index of an oil is calculated from its viscosities in centistokes at 40 °C and 100 °C.

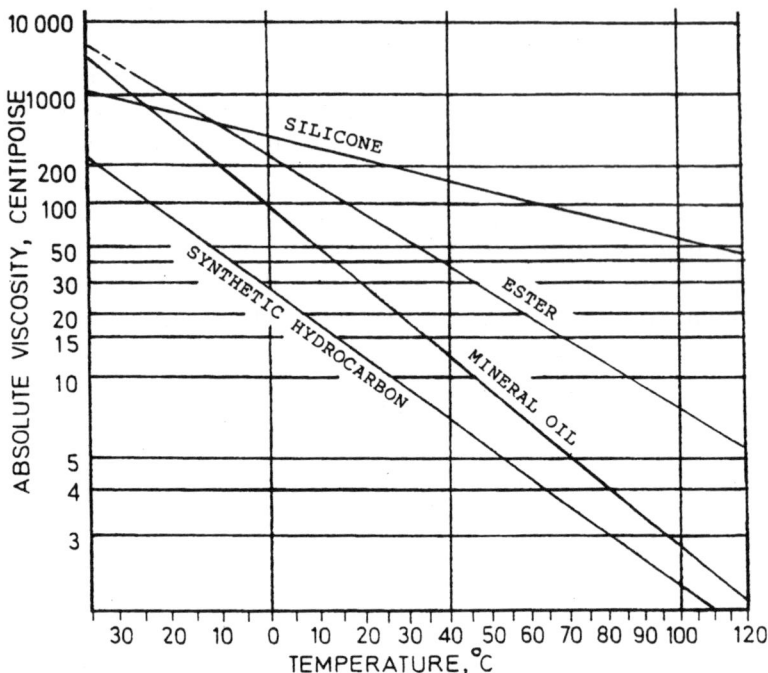

Figure 3.2 Variation of viscosity with temperature

$$VI = \frac{100(L - U)}{(L - H)}$$

where

U is the viscosity of the oil sample in centistokes at 40 °C

L is the viscosity at 40 °C of an oil of 0 viscosity index having the same viscosity at 100 °C as the sample

H is the viscosity at 40 °C of an oil of 100 viscosity index having the same viscosity at 100 °C as the sample

The normal range of VI is from 0 to 100, but the measurement range is often extended to much higher values. If an oil has a Viscosity Index of 0, its viscosity changes rapidly with change in temperature. If it has a Viscosity Index of 100, its viscosity changes less with change in temperature.

Among the mineral oils, the highest natural viscosity index is obtained with paraffinic oils and the lowest with naphthenic oils. Oils having a natural VI above about eighty are called HVI (high viscosity index); those with a natural VI below about thirty are LVI (low viscosity index) while the intermediate oils are MVI (medium viscosity index). Highly refined mineral oils may have VI as high as 130, and are called VHVI (very high viscosity index). Severely refined oils may have VI as high as 150 and are called UHVI (ultra-high viscosity index) oils.

Theoretically, a high viscosity index is only useful if a machine must operate over a wide range of temperatures. In practice there is a tendency to use only high-viscosity index base oils for manufacturing the better quality oils, so that an oil with a low viscosity index may also be inferior in other respects.

It is possible to increase the VI of an oil by dissolving in it a polymeric 'VI improver', such as a polybutene or a polyacrylic. In the past most multigrade engine oils – such as 20W/50 – have been treated in this way. The VI of modern VHVI or UHVI oils may be high enough to provide multigrade properties without the use of any VI improver, but the widest grade ranges – such as 10W/40 – must still contain VI improver.

Before leaving the subject of viscosity index improvers one cautionary note should perhaps be given. When an oil contains a polymeric viscosity index improver, its apparent viscosity will depend on the rate at which it is being 'sheared' or made to flow. Figure 3.3 shows the way in which the viscosity of an oil changes with shear rate.

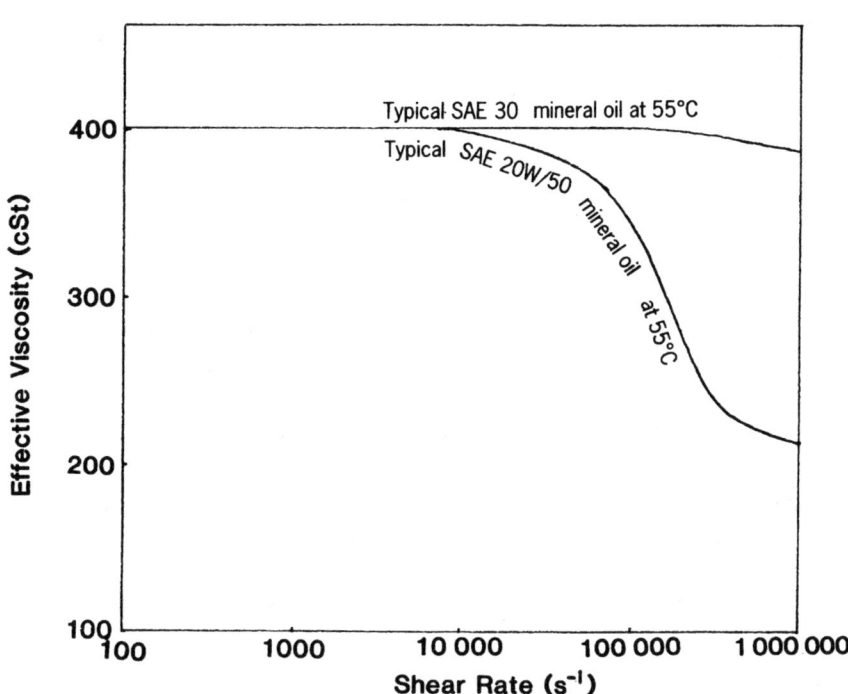

Figure 3.3 Effect of shear rate on viscosity

In the standard viscosity measurements used for oils the shear rate is low. This means that the quoted viscosity, VI and SAE Grade represent the behaviour of the oil at low shear rate. In a bearing, however, the shear rate may be very high, so the effective viscosity of a polymer-containing oil may be much lower. The high quoted viscosity applies when the oil moves slowly, for example when passing through feed pipes. One important effect of this is that, on shutting down an engine, the VI-improved oil drains very slowly from the hot bearings, pistons, cylinder walls, and so on. Consequently, on re-starting the engine a better lubricant film is present.

There is, however, a disadvantage. If a multigrade oil is to have a viscosity at $-15\,^{\circ}$C equivalent to an SAE 15W oil, then the base oil must have a lower viscosity, since the viscosity will be increased by the polymer. When the oil is then sheared rapidly in a high-speed bearing, it may behave as if its viscosity is the same as that of the base oil. In other words, in a fast bearing at $100\,^{\circ}$C the 15W/40 oil may be behaving like an SAE 10W oil rather than as an SAE 40 oil.

In a car engine, for which these oils are intended, this will almost certainly cause no problems, since multigrade oils are intensively tested in car engines. If it is proposed, however, to use a multigrade engine oil in some other type of application, this effect of high shear rate in the bearing must be taken into account. It could produce the opposite result to the one intended.

Early VI improvers tended to become ineffective at shear rates above $10^4\,\text{s}^{-1}$; such shear rates are commonly found in plain bearings. VI improvers are now available which retain their effectiveness at shear rates as high as $10^6\,\text{s}^{-1}$, so that the problem of shear thinning is less common. It is not clear, however, whether the early VI improvers have yet been completely replaced by the more shear-stable types.

At low temperatures the viscosity of a mineral oil may increase very rapidly, as the various liquid molecules approach their freezing points. With paraffinic mineral oils there is also a tendency for waxy molecules to separate out and to thicken the oil like a jelly or grease. This tendency can be reduced by dewaxing the oil, or by adding a very small quantity of a polymer called a 'pour point depressant'.

Certain other viscosity units such as Redwood No. 1 Seconds, Saybolt Universal Seconds, and Degrees Engler are still used. Their relationships to centistokes are explained in the Appendix at the end of the chapter. In addition the SAE (Society of Automotive Engineers) viscosity grade system is so widely used that it is reproduced in Table 3A.1 in the Appendix.

In practice the selection of oil viscosity is often not too critical. If an oil is used with a viscosity higher than the selection procedures suggest, the effect is to increase the resistance to movement in the bearing because of the higher viscous friction or drag. The system consequently becomes hotter, and as the temperature goes up the oil viscosity goes down. The end result is that the system settles down at a slightly higher temperature and a slightly higher oil viscosity. This result is only likely to be unacceptable if the temperature is already high, or the available power is limited.

3.4 Boundary lubrication

Boundary lubrication was introduced in Section 1.5 of Chapter 1. The need for boundary lubrication arises when two surfaces are not completely separated by an oil film, and contact takes place between asperities on the surfaces. This can happen if the oil viscosity is too low, the bearing speed is too low, or the bearing load is too high. It can also happen if, for some reason, the bearing becomes partially starved of lubricant, if the surfaces are not designed to produce a converging wedge, such as in a spool valve, a flat-face sliding valve, or a machine-tool slideway.

Boundary lubrication therefore becomes important whenever the lubricant film thickness is similar to, or less than, the combination of the surface roughnesses. The type of additive needed to provide boundary lubrication depends on the conditions present in the particular system.

The rubbing contact may be fairly gentle, such as in a bearing which makes frequent starts and stops, or in a valve which has no wedge action but is only lightly loaded. Under those conditions mild wear takes place entirely within the surface oxide, and many oils will provide adequate boundary lubrication, especially if they contain naphthenes, sulphur, aromatics, or asphaltenes. If not, then a mild adsorbed or chemisorbed 'anti-wear additive' will probably be all that is needed. Table 3.2 lists a number of different anti-wear additives, and most high-quality lubricating or hydraulic oils are likely to contain between 0.1 and 0.5 percent of such an additive.

In other cases the rubbing contact may be severe, either because the load is very high, or because under moderate load the sliding speed is

Table 3.2 Some anti-wear additives

Ethyl stearate
Stearic acid
Tri-*para*-cresyl phosphate
Tri-xylyl phosphate
Rapeseed oil
Methyl stearate
Zinc diethyl dithiophosphate
Dilauryl phosphate

high. Examples are metalworking techniques where the contact pressures are high enough to deform or cut metals, and hypoid gearboxes where the contact pressures are high and the amount of sliding is also high. In such cases the surface oxide may be completely removed, so that a more powerful 'extreme-pressure additive' will be needed. Table 3.3 lists a number of different extreme pressure additives.

One special type of boundary lubrication problem arises in worm gears, where the need is more for friction reduction than for wear reduction. This friction reduction can be obtained by using small quantities of natural oils (vegetable oils or animals fats) in a mineral oil. Such oils are often called 'compounded oils', and are also used in other types of gearboxes. They have also been widely used in the axle boxes of railway wagons when these employed plain bearings and there was a danger of too much frictional heating. Most railway axles now use roller bearings, and the need for compounded oils is very much less. A more effective modern solution to the worm gear requirement is to use polyglycols instead of mineral oils, but care must be taken to ensure that the two are not mixed, as they are incompatible.

There are, of course, some corresponding disadvantages to the use of boundary additives. They add to the cost of lubricants, because they are costly, and because dispersing them in oil involves technical problems and additional effort. They are also less stable, so that they may reduce the service life of the oil.

The most important disadvantage of extreme pressure additives is that they react with the bearing surfaces, so producing some degree of corrosion. The best of them only react while they are actually preventing bearing seizure, and are inert the rest of the time. Some of the more powerful extreme pressure additives used in metalcutting fluids are more highly corrosive.

Table 3.3 Some extreme pressure additives

Sulphurized mineral oil
Sulphurized fatty oils
Sulphurized oleic acid
Cetyl chloride
Phosphosulphurized fatty oils
Mercaptobenzothiazole
Sulphurized terpenes
Alkyl and aryl polysulphides
Chlornaphtha xanthate
Chlorinated paraffinic oils
Chlorinated paraffin wax
Molybdenum disulphide

These problems are mentioned again later. In general it can be said that small concentrations of mild anti-wear additives are probably harmless. Ideally, however, no boundary lubricant additives should be used where they are not needed, or in higher concentrations than are needed.

3.5 Oil deterioration

It was said earlier that, provided the viscosity and boundary lubrication properties of an oil are satisfactory, the oil will perform properly. However, the properties of the oil will not remain unchanged indefinitely, and a major factor in oil performance is to maintain the required properties for the necessary period of service.

3.5.1 *Oxidation*

The most important form of chemical breakdown of oils and additives is oxidation. Most chemical substances react more or less slowly with oxygen, and because lubricating oils usually operate in contact with air, the various chemicals present react slowly – but continuously – with the oxygen in the air. The rate of oxidation varies considerably between different compounds, but as a rule they follow this order of decreasing resistance to oxidation.

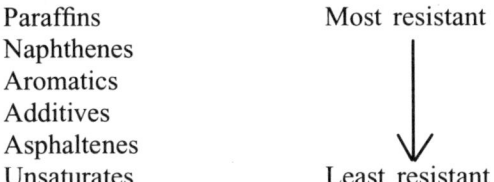

Paraffins   Most resistant
Naphthenes
Aromatics
Additives
Asphaltenes
Unsaturates   Least resistant

The effects of oxidation are to produce aldehydes and acidic compounds, which can cause corrosion, together with an increase in viscosity, and eventually lacquering, tarry deposits, and insoluble oxidation products. All these effects are undesirable, so that a major objective in the refining of mineral oils is to remove the unsaturates, aromatics, and asphaltenes, all of which reduce the oxidation resistance of the oils.

It is unfortunate that the asphaltenes, unsaturates, and – to a lesser extent – naphthenes and aromatics, which reduce oxidation resistance, also tend to improve boundary lubrication. It follows that an oil which has been only mildly refined will have poor oxidation resistance and relatively good boundary lubrication. Its oxidation resistance can be improved by the use of anti-oxidants, but the effect of boundary lubrication additives may be limited.

On the other hand, a severely refined oil will have good oxidation resistance but poor boundary lubrication. Its boundary lubrication can be improved considerably by the use of additives, but anti-oxidants will be less effective. Refining is therefore a compromise. The severity of refining will be chosen to give the optimum balance of properties for a particular grade of lubricant. Overall, the modern tendency is to use a high level of refining to produce a base oil with good oxidation resistance and a high viscosity index. The other required properties are then obtained by the use of additives.

In spite of the good oxidation resistance obtained with modern base oils, most mineral oil lubricants contain anti-oxidants. They are more widely used than any other class of additive; even oils which are loosely described as 'non-additive' will, in fact, usually contain a small amount of anti-oxidant.

Anti-oxidants are additives which reduce the harmful effects of oxidation. The oxidation of most mineral oil components and additives proceeds by a sort of chain reaction, involving organic peroxides. Most anti-oxidants act by reacting with an organic peroxide so blocking the chain. Some of the commoner anti-oxidants are listed in Table 3.4.

Table 3.4   Some examples of anti-oxidants

Type	Example
Metal organophosphates	Zinc diethyl dithiophosphate
Amines	Phenothiazine
	N-phenyl-naphthylamine
	Diphenylamine
Hindered phenols	2,6-di-*tert*-butyl-4-methylphenol
Organic phosphites	Tri-*n*-butyl phosphite
Organometallics	Zinc-di-*n*-butyldithiocarbamate

Due to this very specific blocking action of anti-oxidants, they continue to prevent oxidation as long as they are present, even in very small concentration. Ultimately, however, they become completely used up, and there is nothing left to inhibit oxidation. The oil then starts to oxidize rapidly. The effect is shown schematically in Fig. 3.4. The important thing to remember is that oxidation is very slow until all the anti-oxidant is used up; it may then take place very quickly. Ideally, an oil should be changed just before the anti-oxidant is used up; in some cases it is possible to arrange this.

3.5.2 Thermal stability

Even if there is no oxygen present, an oil cannot be heated above a certain temperature without starting to decompose. This type of

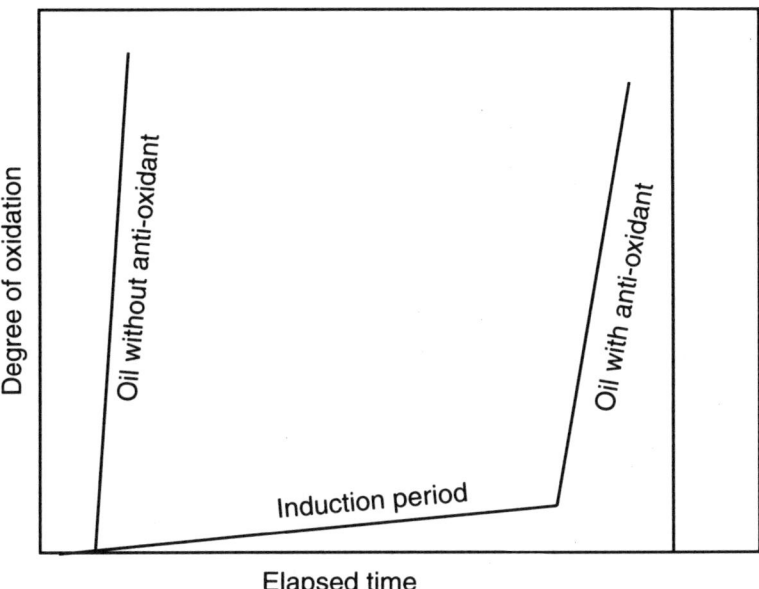

Figure 3.4   Oxidation of a mineral oil

breakdown is called thermal degradation, and it causes discoloration of the oil and a change in viscosity. The thermal stability cannot be improved by the use of additives. It can be improved in the refining process by removing those same compounds which lower the oxidation resistance.

Figure 3.5 gives an idea of the temperature limits for the use of mineral oils. The rates of oxidation or thermal degradation increase with temperature, so that higher temperatures can be tolerated if the required life is shorter. There is therefore no absolute temperature above which an oil cannot be used, unless it boils.

The lower temperature limit shown in Fig. 3.5, is imposed by the increase in viscosity as the temperature decreases. Again, there is no absolute limit because the maximum acceptable viscosity depends on the power which is available to move the lubricated parts. However, a useful guide to the minimum possible temperature is the pour point – the temperature below which the oil will not pour.

From all this it can be seen that for long service at high temperatures highly refined oils with anti-oxidants should be used. For lower temperatures less highly refined oils may have some advantages, as well as being cheaper.

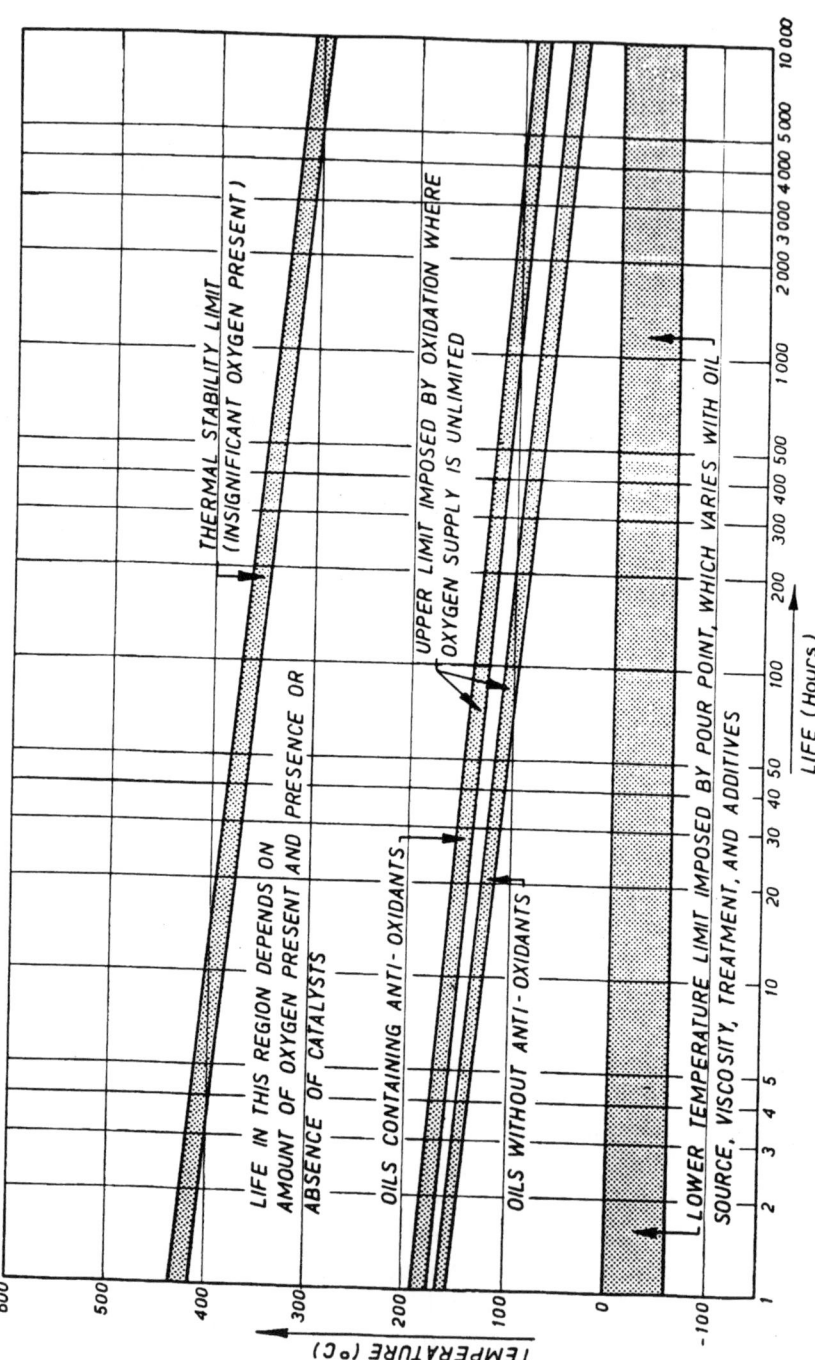

Figure 3.5 Temperature limits for use of mineral oils

3.6 Contamination

Apart from chemical breakdown, oils may deteriorate in service because of contamination. There are many different possible contaminants, including:

— Water from condensation or combustion
— Unburned fuel in an engine
— Wear debris
— Dust from the atmosphere
— Process liquids
— Chemicals in chemical plant
— Soot from faulty fuel combustion
— Breakdown products of the base oil
— Corrosion products
— Breakdown products of additives
— Organic debris from microbiological attack

Not only are all these contaminants undesirable in themselves, but two or more of them may act together to cause more damage than they would separately. Any of them can cause further deterioration and more contamination.

Dust, corrosion products, and wear debris can increase wear and thus produce more wear debris.

Acidic breakdown products and water can produce a corrosive mixture and thus generate more corrosion products.

Acidic breakdown products and water can form a surface-active mixture which will emulsify with the oil and block feed holes and filters.

The same emulsion makes an ideal medium for microbiological growth (bacteria and fungi) which feed on the oil itself.

Soot and the debris from microbiological growth can block feed pipes and filters.

It is therefore important to remove all types of contaminant as thoroughly as possible, and even then contamination may sometimes make it necessary to change the oil sooner than would be needed because of oil breakdown.

There are several techniques available for coping with contamination problems.

(a) *Preventing entry of contaminants*
     Contaminants such as atmospheric dust and moisture, chemicals, and process liquids can be controlled by efficient sealing or by filtration of air supplies.
(b) *Removal of contaminants*
     In a circulating oil system, solid or liquid contaminants can be

removed by filtration or centrifuging. In general, filtration is more effective for solid particles, and centrifuging for liquids. Large ferrous particles can be removed by a magnetic plug.

(c) *Dispersing contaminants*

Certain solid contaminants, such as soot and oil breakdown products, tend to accumulate together and form large particles which can block oilways or filters. It may sometimes be preferable to keep the contaminants dispersed in the oil by using dispersant or detergent additives, so that the particles are too fine to cause any problems. In general, however, dispersion of contaminants is less satisfactory than removing them.

(d) *Neutralization*

Acidic products from oil breakdown or from burning of sulphur-containing fuels may be neutralized by basic additives, such as calcium compounds, in order to prevent corrosion.

Eventually, however effective the cleaning techniques, a time comes when the contamination interferes with the quality of the lubrication. This is one of the main factors which make oil changes necessary.

3.7 Compatibility

Lubricants must always be compatible with the other materials present in the system. In other words the lubricants and the other materials should not have any undesirable effect on each other.

Some of the most important problems arise between lubricating oils and rubbers, which must be used for seals and flexible hoses. Other problems can arise between oils and plastics, adhesives, and paints. Some synthetic oils will even attack metals.

Less highly refined mineral oils – or those which are oxidized – and some of the common additives will attack certain metals, especially when hot. Ordinary corrosion is considered later, but there is also a form of attack in which a metal, especially copper, can be dissolved in an oil and re-deposited in some other part of the system. For this reason copper, brass, and bronzes should be avoided if possible in situations where they would come into contact with hot oil.

Compatibility problems with synthetic oils are much more difficult, and will be discussed later.

3.8 Corrosion

Clean mineral oils are mainly non-corrosive, and will in fact give very good protection against corrosion by atmospheric moisture. Oxidation products and some additives will attack steel and bearing metals such as copper, bronze, tin, and aluminium – especially if there is water present and if the oil is hot. Water is often present, and an oil does little to prevent corrosion by free water.

Oils which are to be used in situations where there could be a particular risk of corrosion often have a corrosion inhibitor added to them. Table 3.5 lists some types of corrosion or rust inhibitor.

Table 3.5   Some examples of corrosion and rust inhibitors
Zinc diethyldithiophosphate
Zinc diethyldithiocarbamate
Trialkyl phosphites
Sulphurized terpenes
Calcium or barium sulphonates

3.9   Synthetic and natural oils and emulsions

So far this chapter has concentrated mainly on mineral oils, as these are by far the most widely used oils. They have the advantages of relative cheapness, high stability, availability in a wide range of viscosities, and good boundary lubrication. Any liquid can be used as a lubricant under certain conditions, and such varied liquids as liquid sodium or potassium, sea-water, molten glass, silicones, and animal fats have all been used successfully.

Some of the reasons for using a different base oil in place of a mineral oil were briefly considered in Chapter 2. They include:

– Temperature too high for mineral oil
– Temperature too low for mineral oil
– Lower flammability needed
– Compatibility problems, e.g., with natural rubber
– Contamination problems, e.g., with food

Figure 3.6 shows approximate temperature limits for some synthetic oils.

Many of the alternative types of oil are synthetic, or in other words they are manufactured from various feedstocks by chemical processes. They are often described as synthetic oils, but the term is always ambiguous and potentially misleading and dangerous, because different types of synthetic oil are very different from each other in their performance and properties. It follows that whenever a synthetic oil is mentioned, it should be made clear what type of synthetic it is. Particular problems which can arise if the type of synthetic oil is not identified include compatibility with non-metallic materials and mixing with other oils.

In spite of this danger, suppliers of synthetic oils have consistently

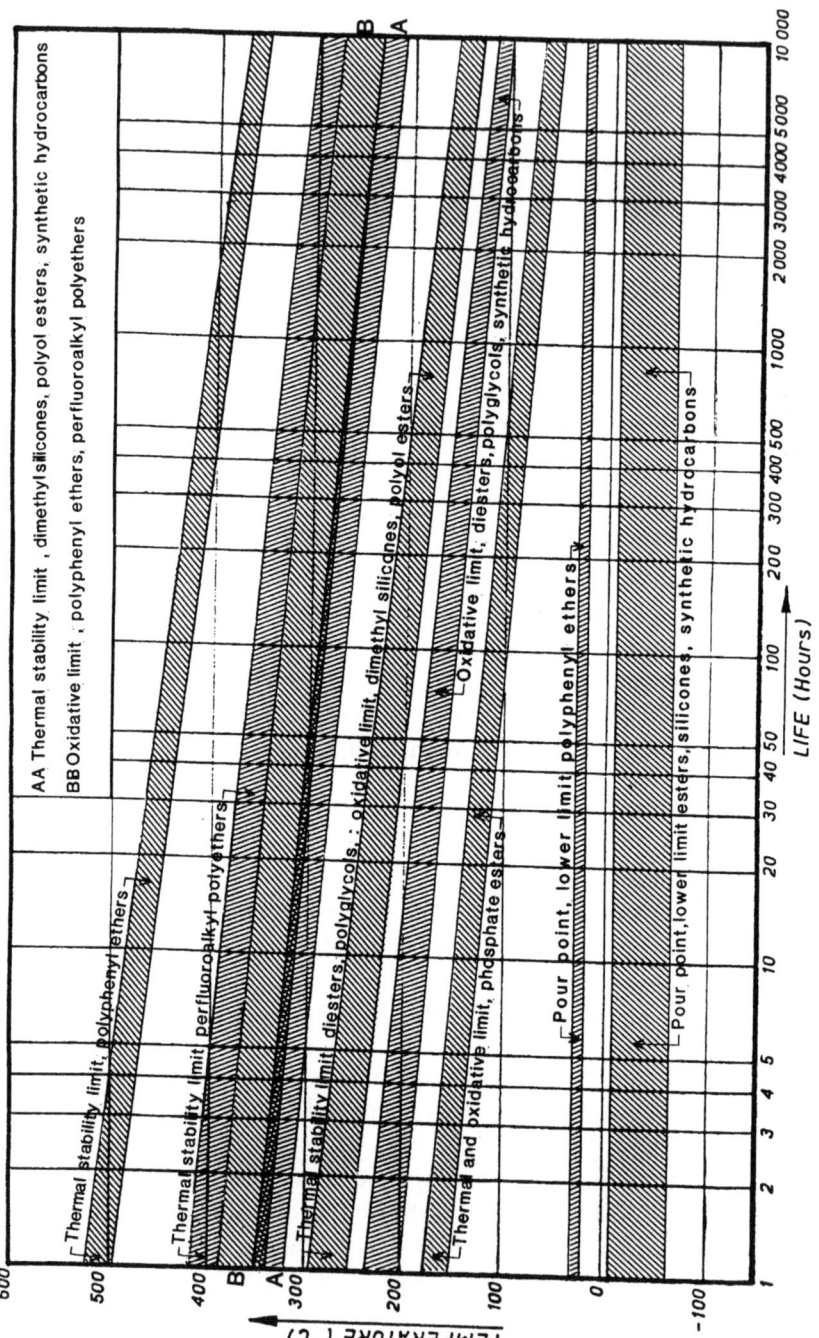

Figure 3.6 Temperature limits for some synthetic oils

failed to identify the types of synthetic oil used in their products. The reason for this is not at all clear. Lubricant suppliers are usually prepared to specify whether their mineral oils are paraffinic or naphthenic, although this is rarely hazardous. However, they will often not provide the much more critical information about the types of synthetic oil present.

The widely-used synthetic oils include hydrocarbons, di-esters, polyol esters, phosphate esters, silicones, polyglycols, and – to a much smaller extent – polyphenyl ethers and perfluoroalkylpolyethers. Silicones are normally referred to as silicones, and not as synthetics. Phosphate esters are also often, but not always, specified as such. Polyglycols may be described specifically enough to avoid confusion. The biggest risk of confusion, therefore, arises with the synthetic hydrocarbons, di-esters, and polyol esters. These differ so much in their effects on rubber seals and hoses that any use of the wrong oil could be disastrous.

3.9.1 Synthetic hydrocarbons

There are several classes of synthetic hydrocarbon. The most important are branched-chain paraffins generally similar to the one in Fig. 3.1(b) and are known as polyalphaolefines (PAO). Their properties and performance are very similar to some of the most highly refined mineral oils; for practical purposes they can be considered almost indistinguishable from them.

This highlights the whole problem raised by the vague use of the general word 'synthetic'. Polyalphaolefines are one of the most important classes of synthetic oil, yet they resemble mineral oils, the most important non-synthetic, far more closely than they resemble any of the other synthetic oils.

The major use of polyalphaolefines is in engine oils for motor vehicles. Their inherent oxidation resistance is quite good and their boundary lubrication not very good, like those of the highly refined mineral oils. With the proper additives they can be considerably improved, and they then have a much better life and performance than conventional mineral oils, especially at higher temperatures.

Also, like the very highly refined mineral oils, they tend to shrink rubber seals and hoses, or at best not to swell them. It is usual to blend with them a small proportion of a synthetic ester to give the small rubber swell which is desirable for good sealing.

3.9.2 Di-esters and polyol esters

In terms of the total volume used, di-esters and polyol esters comprise the most important class of synthetic oils, as they are almost universally used for aircraft gas turbine engine lubrication. The earliest jet engines

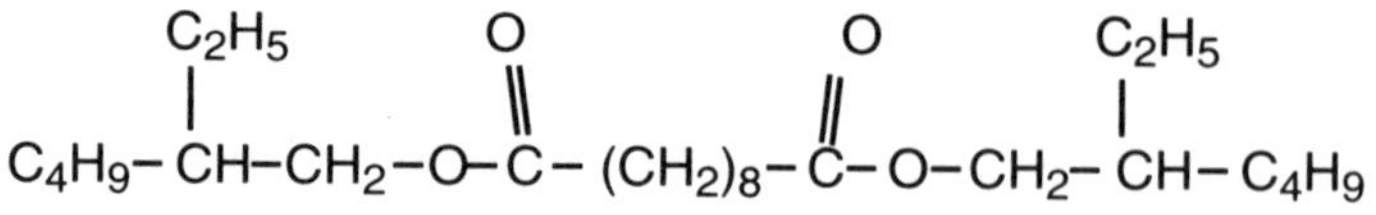

$$C_4H_9-\overset{\overset{\displaystyle C_2H_5}{|}}{CH}-CH_2-O-\overset{\overset{\displaystyle O}{\|}}{C}-(CH_2)_8-\overset{\overset{\displaystyle O}{\|}}{C}-O-CH_2-\overset{\overset{\displaystyle C_2H_5}{|}}{CH}-C_4H_9$$

Typical Di-Ester (di-2-ethylhexyl Sebacate)

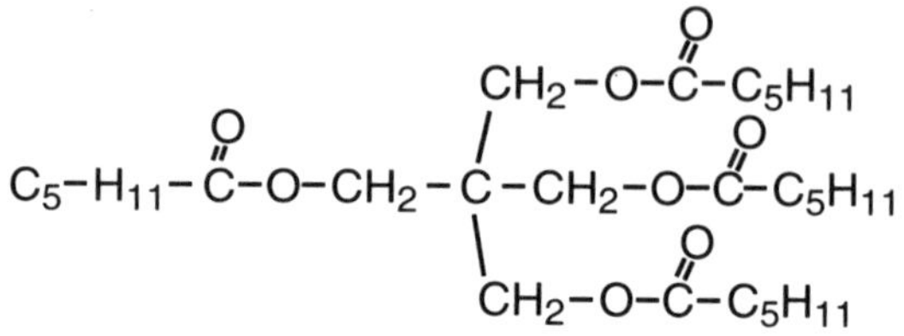

$$C_5-H_{11}-\overset{\overset{\displaystyle O}{\|}}{C}-O-CH_2-\overset{\overset{\displaystyle CH_2-O-\overset{\overset{\displaystyle O}{\|}}{C}-C_5H_{11}}{|}}{\underset{\displaystyle CH_2-O-\overset{\overset{\displaystyle O}{\|}}{C}-C_5H_{11}}{C}}-CH_2-O-\overset{\overset{\displaystyle O}{\|}}{C}-C_5H_{11}$$

Pentaerythritol Tetrahexanoate

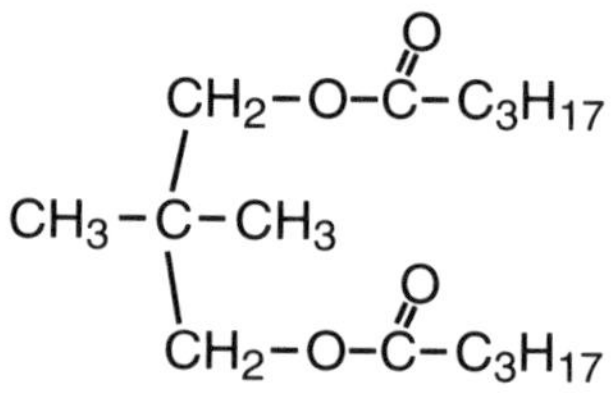

$$CH_3-\overset{\overset{\displaystyle CH_2-O-\overset{\overset{\displaystyle O}{\|}}{C}-C_3H_{17}}{|}}{\underset{\displaystyle CH_2-O-\overset{\overset{\displaystyle O}{\|}}{C}-C_3H_{17}}{C}}-CH_3$$

Neopentyl Glycol Dinonanoate

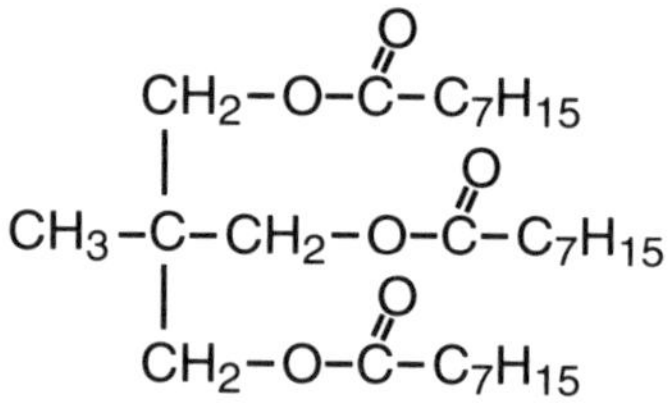

$$CH_3-\overset{\overset{\displaystyle CH_2-O-\overset{\overset{\displaystyle O}{\|}}{C}-C_7H_{15}}{|}}{\underset{\displaystyle CH_2-O-\overset{\overset{\displaystyle O}{\|}}{C}-C_7H_{15}}{C}}-CH_2-O-\overset{\overset{\displaystyle O}{\|}}{C}-C_7H_{15}$$

Trimethylolethane Trioctanoate

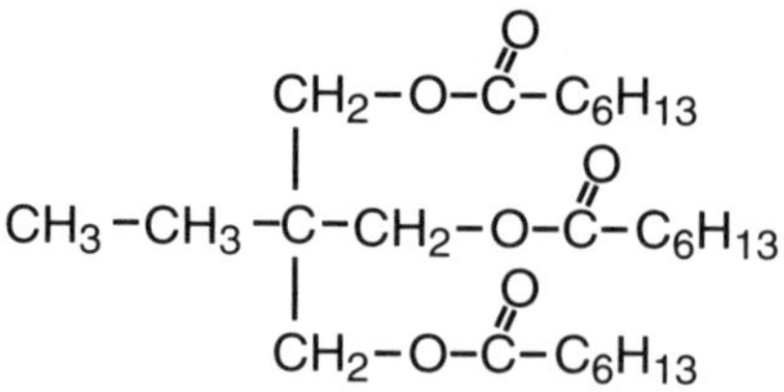

$$CH_3-CH_3-\overset{\overset{\displaystyle CH_2-O-\overset{\overset{\displaystyle O}{\|}}{C}-C_6H_{13}}{|}}{\underset{\displaystyle CH_2-O-\overset{\overset{\displaystyle O}{\|}}{C}-C_6H_{13}}{C}}-CH_2-O-\overset{\overset{\displaystyle O}{\|}}{C}-C_6H_{13}$$

Trimethylolpropane Triheptanoate

Figure 3.7   Typical di-ester and polyol esters

were lubricated with mineral oils, but some of the temperatures were too high for reliable operation. Many possible alternative oils were tested, but the best were found to be di-esters. By about 1960 almost all jet engines were lubricated with di-esters.

Di-esters are a class of 'carboxylic esters', containing the carboxylic group $-0-C=0$. A typical example is di-2-ethylhexyl sebacate, shown in Fig. 3.7.

As aircraft gas turbine engines became more powerful and aircraft speeds higher, the engine operating temperatures rose until di-esters were also inadequate. A new class of carboxylic ester oils was then developed with better thermal stability. These are called 'polyol esters' or sometimes 'complex esters', and a typical example is trimethylol-propane triheptanoate. Figure 3.7 shows four polyol esters used in lubricants.

Di-esters are used as base oils for high-temperature greases and as components of synthetic vehicle engine oils. They are also becoming more widely used for high-temperature applications in industry, such as hot-rolling oils in steel rolling. They can be used at higher temperatures than mineral oils, have very good lubricating characteristics, and are readily available in a variety of viscosities.

Polyol esters are now extensively used in aircraft jet engines and are readily available in a few different viscosity grades. Apart from their good high-temperature properties, they also have very good lubricating characteristics, but they are not yet widely used outside aviation.

3.9.3 *Phosphate esters*

These oils are the esters which are produced by reaction between alcohols and phosphoric acid. They are quite widely used for their outstanding fire resistance, especially in high fire-risk situations such as aircraft hydraulic systems, coal mines, and hot metal-processing. They are chemically similar to some of the best anti-wear additives, and consequently have excellent boundary lubricating behaviour. They are available in various viscosity grades and are used as base oils for fire-resistant greases. Their main disadvantages are poor thermal stability, which limits them to temperatures not much higher than 100 °C, and their powerful solvent action on many paints, plastics, and rubbers.

3.9.4 *Silicones*

Silicones are polymeric substances based on a chain skeleton containing alternate silicon and oxygen atoms. They are available in a very wide range of viscosity grades, from those which are less viscous than gasoline to those which are as viscous as asphalt. They are generally used for their good high-temperature stability, which enables them to be used over 200 °C. Their other advantages include chemical inertness, water repulsion, and electrical insulation. They are widely used as base oils for high-temperature greases. Their main disadvantage is that they are poor boundary lubricants, especially for steel against steel. The phenyl silicones are better in this respect, and the chlorinated and fluorinated silicones even better, but these are not used very widely.

3.9.5 Chlorinated biphenyls

The chlorinated biphenyls have very good fire resistance and chemical inertness, but are mediocre boundary lubricants. They are used mainly in hydraulic systems and compressors, especially in situations where there is a serious fire risk.

3.9.6 Polyglycols

Polyglycols are long-chain polymeric liquids. They are stable to about 200 °C. Their greatest advantage is that, when heated above this temperature, they decompose cleanly without producing any undesirable decomposition products.

There are two different types, one of which is water-soluble. This type is used for lubrication of pumps handling petroleum products or other solvents, since it resists washing out by the solvents.

3.9.7 Fluorinated ethers and fluorocarbons

These oils have the highest thermal stability and chemical inertness of any oils, and can be used continuously at temperatures over 300 °C. Like the silicones, however, they are poor boundary lubricants, and at present they are extremely expensive.

3.9.8 Polyphenyl ethers

Polyphenyl ethers also have excellent thermal stability and can be used continuously up to 300 °C. They are better boundary lubricants than the silicones and fluorinated ethers, but are still only mediocre. One important disadvantage is that they are solid at normal temperatures.

3.9.9 Natural oils

These include vegetable oils and animal fats, which were for many centuries the only commonly used oils. They are usually excellent boundary lubricants, but they are much less stable than mineral oils, and tend to break down to give sticky deposits. Rapeseed oil is still used by itself or as an additive to mineral oils to give improved boundary lubrication, and is used in some hot metalforming processes because it does not cause carburization of steel. Castor oil is used as a hydraulic fluid in a few aircraft and is still available for some older aircraft rotary engines and racing car engines. It has very good boundary lubrication and excellent compatibility with natural rubber, but its use for engines and hydraulic systems is now very small.

However, one advantage of natural oils and fats is that they are readily bio-degradable. A recent development is an increase in use of castor oil and palm oil in manufacture of bio-degradable and food industry grade lubricants. Bio-degradability is considered highly desirable in the increasing concern for protection of the environment. For the first time in many years an engine oil based on natural oils was recently developed for automotive use.

3.9.10 *Emulsions*

The oils described so far have been true solutions or pure liquids. This means that their molecules are fully and uniformly mixed together and will not separate out on standing, or even on powerful centrifuging.

If some water is simply added to a mineral oil, or an ester or a silicone, it will form a separate layer because it is insoluble in them. It is possible, however, to disperse water in an oil, or an oil in water, in the form of tiny droplets; the resulting dispersion is called an emulsion. To prevent the droplets from combining together and separating out, the emulsion must be stabilized. This is usually done by the use of a surface-active material such as a soap or detergent. The molecules of the surface-active material concentrate at the surfaces of the dispersed droplets. Because of their electronic structure they produce an electrical double layer in which all the droplets carry the same charge, either positive or negative, while the liquid around them carries the opposite charge. Because all the droplets have the same charge they repel each other, and cannot combine or coalesce.

An emulsion therefore consists of small droplets of one liquid uniformly dispersed in a second liquid, neither liquid being soluble in the other. The emulsion is electrically stabilized by some type of surface-active substance.

In practice, the emulsions used in lubrication usually contain water and a mineral oil as the two mutually insoluble liquids. The earlier types always consisted of oil dispersed in water. In recent years, however, emulsions of water in oil have been developed; these are known as invert emulsions. Since the oil is the continuous phase in invert emulsions, their lubricating properties are generally those of the oil, and are very good. The water droplets help to prevent fire by their powerful cooling effect, and by quenching fires with steam. However, they have relatively little effect on lubrication except for limiting the temperature of use, usually to less than about 80°–85 °C.

3.9.11 *Liquefied glasses, metals, and salts*

These materials can all be used as lubricants at temperatures over 300 °C, but they involve handling problems because they are solid at normal temperatures. Liquid glasses are used in some hot metalforming processes, but other applications of all of them are highly specialized.

Appendix

Viscosity units

Viscosity is defined in two different ways and measured in many different ways, but they can all be related to each other. The relationships between four of them can be seen in Figure 3A.1.

Kinematic viscosity

This is the viscosity most often quoted by lubricant manufacturers and users. It is measured by means of a number of different capillary tube

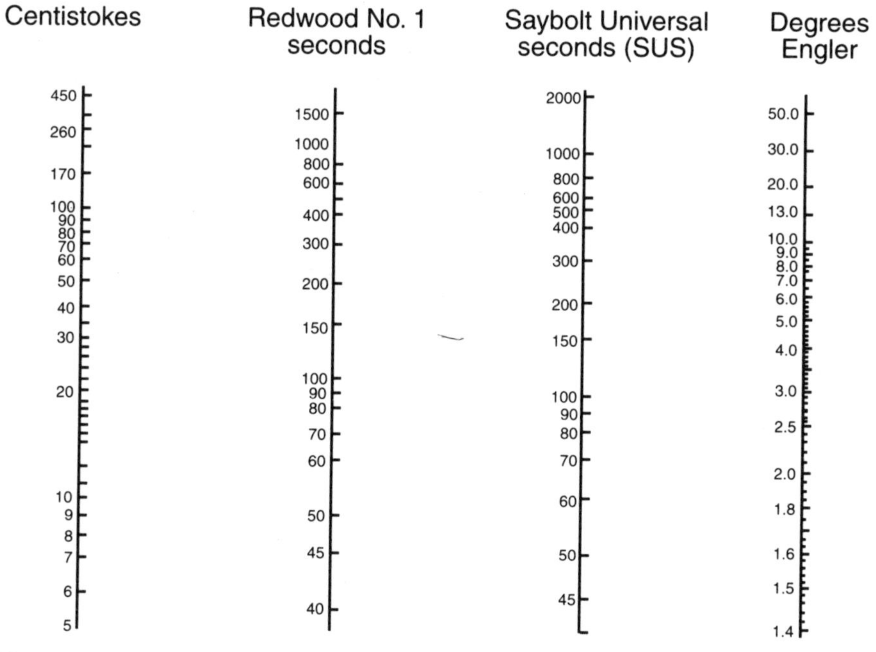

Centistokes	Redwood No. 1 seconds	Saybolt Universal seconds (SUS)	Degrees Engler
450	1500	2000	50.0
260	1000	1000	30.0
170	800	800	20.0
100	600	600	13.0
90	400	500	10.0
80		400	9.0
70	300		8.0
60		300	7.0
50	200		6.0
40		200	5.0
30	150	150	4.0
20	100		3.0
	90	100	2.5
	80	90	
	70	80	2.0
	60	70	
10		60	1.8
9	50		
8			1.6
7	45	50	
6		45	1.5
5	40		1.4

Figure 3A.1   Comparison of viscosity units

devices in which the oil flows through the tube under gravity. (See Chapter 9.)

It is most often measured in Centistokes (cSt). Water at 20 °C has a kinematic viscosity of approximately 1 cSt. The proper SI Unit is the metre^2 per second, and

1 cSt = 1 millimetre^2/second.

Dynamic viscosity

This is the viscosity which has to be used in calculations for bearing design. It is therefore the preferred viscosity for bearing specialists and design engineers. It is equal to the kinematic viscosity of the liquid multiplied by its density at the same temperature, and can be measured directly by certain specialized viscometers.

It is usually expressed in Poise or Centipoise (cP), and because the density of water at 20 °C is close to 1, its dynamic viscosity is also close to 1. The proper SI Unit is the Newton second per metre^2, or Pascal second, and

1 cP = 10^{-3} Newton second/metre^2 = 10^{-3} Nm^{-2}s

Redwood viscosity

Technically Redwood viscosity is not a viscosity but a flow time. It is the number of seconds taken for 50 ml of the oil to flow through a cup-shaped funnel. The method was once an Institute of Petroleum standard method, but has now been discontinued in favour of kinematic viscosity.

Saybolt viscosity

The Saybolt viscosity is also technically a flow time and not a viscosity. It's measured in a similar manner to the Redwood viscosity. The method was once an ASTM standard method, but has now been discontinued in favour of kinematic viscosity.

Engler viscosity

The Engler viscosity is again based on flow time through a cup-shaped funnel. The result, however, is quoted in degrees, depending on the ratio of the flow time for the oil sample to the flow time for water at the same temperature. It is almost exclusively used on the European continent, and is gradually being discontinued in favour of kinematic viscosity.

Apart from the above five, there is one other viscosity classification which is very widely used and must be described. This is the American Society of Automotive Engineers, or SAE, system. Lubricating oils are

Table 3A.1 SAE viscosity grades

(a) 1995 SAE viscosity grades for engine oils

SAE viscosity grade	Viscosity (cP) at temperature (°C)		Viscosity at 100 °C (cSt)	
	Cold cranking (max.)	Pumpability (max. no yield stress)	Minimum	Maximum
0W	3250 at −30	60 000 at −45	3.8	−
5W	3500 at −25	60 000 at −35	3.8	−
10W	3500 at −20	60 000 at −30	4.1	−
15W	3500 at −15	60 000 at −25	5.6	−
20W	4500 at −10	60 000 at −20	5.6	
25W	6000 at −5	60 000 at −15	9.3	−
20	−	−	5.6	< 9.3
30	−	−	9.3	< 12.5
40	−	−	12.5	< 16.3
50	−	−	16.3	< 21.9
60	−	−	21.9	< 26.1

Note: There is also a high-shear viscosity minimum at 150 °C and 10^6 sec^{-1}.

(b) 1991 SAE viscosity grades for axle and manual transmission lubricants

SAE viscosity grade	Maximum temperature (°C) for viscosity of 150 000 cP	Viscosity at 100 °C(cSt)	
		Minimum	Maximum
70W	−55	4.1	−
75W	−40	4.1	−
80W	−26	7.0	−
85W	−12	11.0	−
90	−	13.5	< 24.0
140	−	24.0	< 41.0
250	−	41.0	−

These tables are reproduced with the kind permission of the Society of Automotive Engineers, USA.

classified by the SAE on the basis of their viscosity at low temperature or 100 °C, in accordance with Table 3A.1.

Oils which can be described simply in terms of one of the SAE numbers are called single-grade oils, e.g., SAE 20W or SAE 90. Oils which meet a low temperature requirement, as well as a higher 100 °C requirement, are called multigrade. For example, an engine oil with a viscosity of 3200 cP at −15 °C and 14.0 cSt at 100 °C is a 15W/40 multigrade oil.

Chapter 4

Selection of Lubricating Oil

4.1 The selection process

The criteria for deciding whether to use a lubricating oil in a machine or component have been described in Chapter 2. This chapter will proceed further to explain how to choose the particular type of oil.

The factors to be considered follow from the important features of lubricating oils described in Chapter 3, and include the following stages:

- Selection of the type of base oil
- Choosing the oil viscosity
- Defining the boundary lubrication requirements
- Other required properties and additives

Logically, the type of base oil should be chosen first, since this choice affects the boundary properties and the other properties and additives required. However, well over 90 percent of lubricating oil consumption consists of mineral oils. The factors involved in selecting mineral oil viscosity, boundary lubrication, and other properties will therefore be explained first, followed by the reasons for choosing other types of base oil.

Because of their special situations, engine oils and metalworking fluids are discussed in separate sections.

4.2 Choosing the correct oil viscosity

Although the viscosity is the most important single property of an oil, in many cases the choice is not very critical, and some fairly simple guidelines and examples will be sufficient. It is only for applications where the choice of the right viscosity is critical that the Reynolds equation (p. 7) and the Dowson–Higginson (p. 9) and similar equations need to be used. Such cases are best referred to specialists.

The full range of lubricating oil viscosities runs from about 3.0 cSt to about 1500 cSt. Table 4.1 lists typical viscosity ranges for a number of different types of oil.

Note that it is the viscosity at the operating temperature which is most important. If an instrument is to work satisfactorily at 100 °C, then the oil should have a viscosity of the order of 10 cSt at 100 °C. On the other hand, if the instrument is to work properly at −30 °C, then ideally the oil should have a viscosity of 10 cSt at −30 °C. This means that two very different oils will be needed. The oil for high-temperature use will probably have a viscosity at ordinary room temperature of about 300 cSt. The one for low-temperature use would have a viscosity at room temperature of about 2 cSt, if such an oil could be found.

In Table 3A.1 the viscosity of the SAE W (W for winter) oils is quoted at low temperatures. This is because they are intended for use in cold climates, where the critical requirement is to be able to start the engine at low temperature, even if the use of a low viscosity oil results in some wear at the normal running temperature. On the other hand, the SAE50 oil is intended for use in hot climates where the main problem is not to ensure a low enough viscosity to start the engine, but to ensure a high enough viscosity to give effective lubrication at a high running temperature. Its viscosity is therefore defined at 100 °C. The object of a multigrade oil such as 10W/40 is to try to satisfy both cold starting and hot running requirements.

Table 4.1 Viscosity ranges for various applications

Oil type	Viscosity range (cSt) at operating temperature
Clock or instrument oil	5–20
Sewing machine oil	10–25
Motor oil	10–50
Turbine oil	10–50
'General purpose' household oil	20–50
Hydraulic oil	20–100
Roller bearing oil	10–300
Plain bearing oil	20–1500
Gear oils:	
Low speed spur, helical, bevel	200–800
Medium speed spur, helical, bevel	50–150
High-speed gears	15–100
Hypoid gears	50–600
Worm gears	200–1000
Open gear lubricants	100–50 000

Some of the oil types in Table 4.1 have a wide range of possible viscosities. Table 4.2 shows the most important factors which determine which part of the range is suitable for a particular application.

Many graphical guides to viscosity selection have been published. Figure 4.1 is an example for plain bearings, such as plain journal bearings and tilting pad thrust bearings. Most manufacturers of ball and roller bearings publish similar graphs relating to their own products. Figure 4.2 is a guide for rolling bearings, while Fig. 4.3 is a simple guide for gear oils.

Table 4.2 Factors affecting choice of viscosity

Low viscosity	Intermediate viscosity	High viscosity
High bearing speed ◄──────────────────►		Low bearing speed
Low load ◄──────────────────►		High load
Fully enclosed ◄──────►	Some air access ◄──────►	Well ventilated
Full oil circulation ◄──────►	Splash or drip ◄──────►	No oil feed
Very small bearings ◄──────────────────►		Large bearings

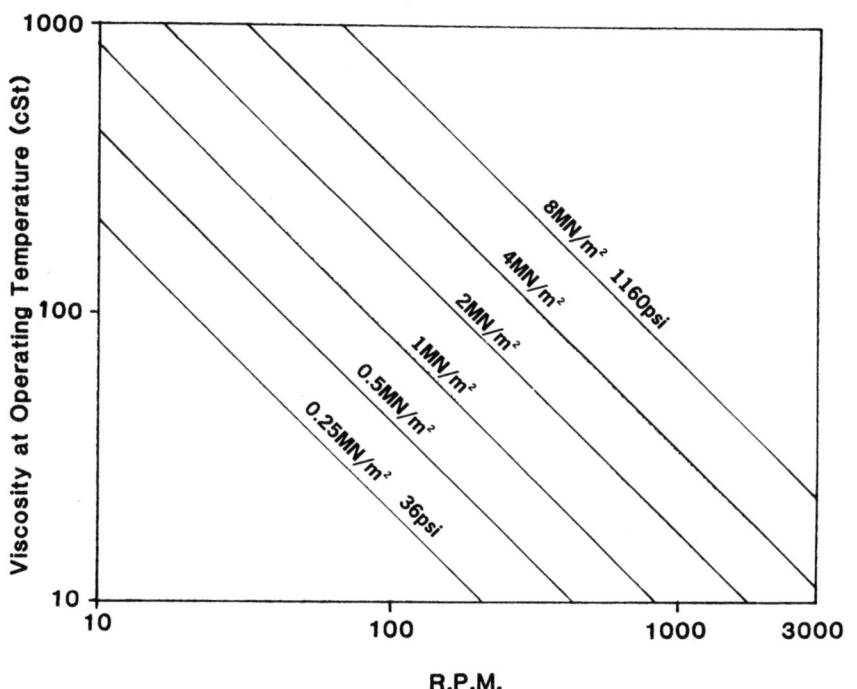

Figure 4.1 Viscosity selection for plain bearings

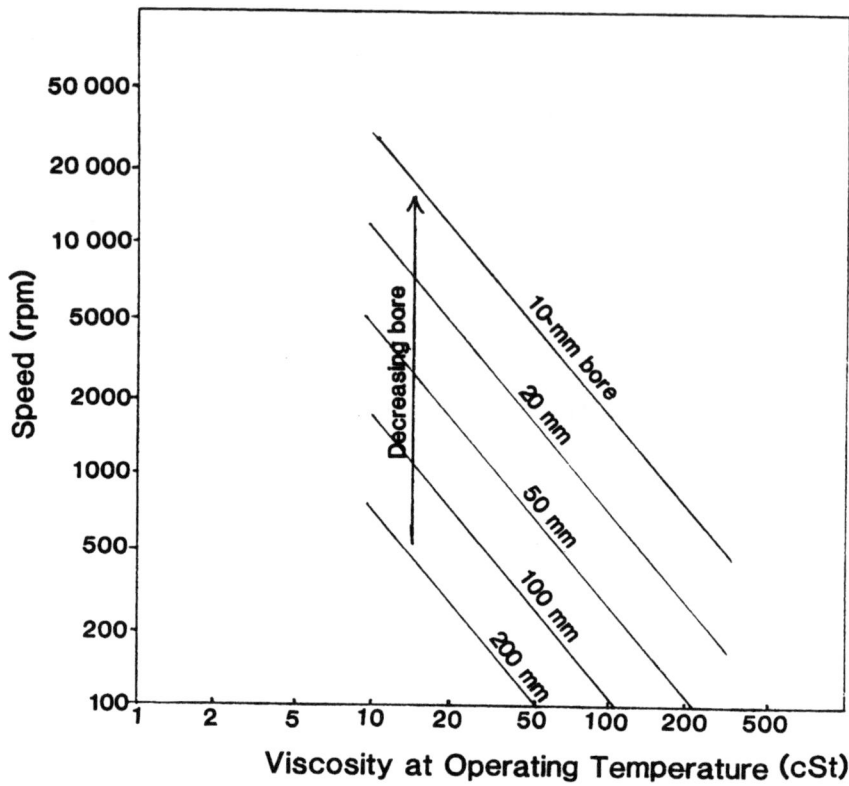

Figure 4.2 Viscosity selection for rolling bearings
Note: an increase in viscosity up to four times the value
given by this figure will give a comparable increase in
bearing life, but may not be acceptable because of
difficulty in pumping or filtration, greater heating, etc.

Having selected the right oil viscosity for the operating temperature, it
must then be adjusted to one of the standard reference temperatures,
such as 40 °C, 50 °C, or 100 °C in order to decide what 'viscosity grade'
is needed. This conversion can only be easily done by using a graph
such as those given in Fig. 3.2 or Fig. 4.4. An alternative method is to
use a chart showing the change of viscosity with temperature. One
useful chart is published in British Standard BS 4231 'Viscosity
Classification for Industrial Liquid Lubricants' (ISO 3448).

As an example, suppose that at an operating temperature of 85 °C a
good quality mineral oil with a viscosity of 11 cSt is needed. The oil
supplier specifies the viscosity of his products at the reference
temperature of 40 °C. From Fig. 4.4 an oil having a viscosity of
11 cSt at 85 °C is oil No 1, whose viscosity at the 40 °C reference

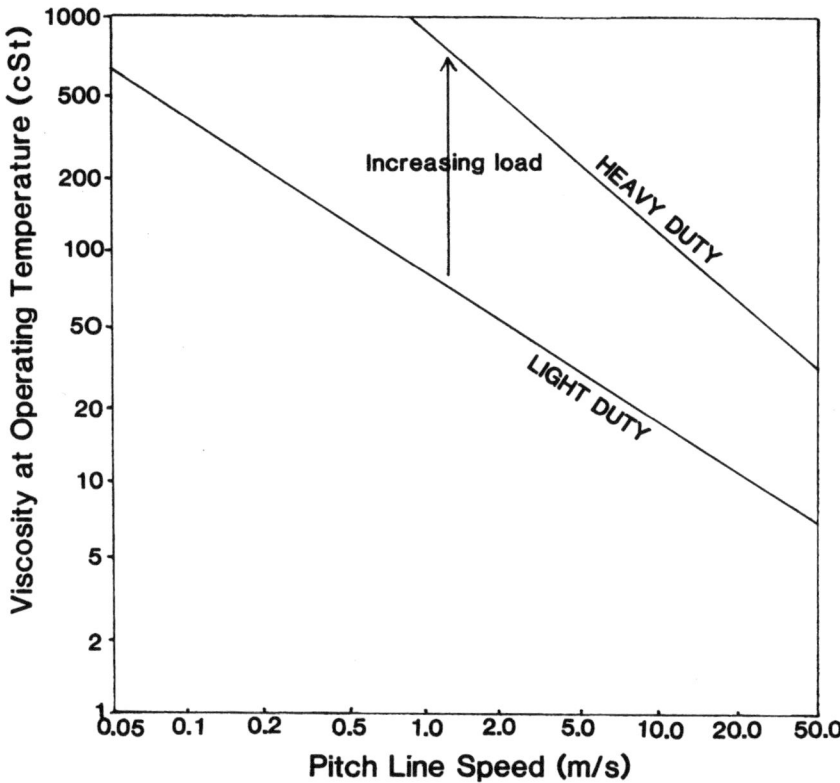

Figure 4.3 Viscosity selection for gears

temperature is 60 cSt. A 60 cSt oil must, therefore, be specified if using the 40 °C reference temperature.

4.3 Boundary lubrication requirements

Most lubricated systems experience mild rubbing under certain circumstances, for example during start-up or run-down, or when new contact surfaces are running-in. For this reason, most lubricating oils contain a small concentration of anti-wear compounds. These may be natural, like the asphaltenes in a mildly-refined oil, or may consist of a small concentration such as 0.1 percent of an anti-wear additive like stearic acid.

There are other systems which have a greater tendency to operate under mild boundary conditions. Hydraulic systems are particularly likely to experience mild wear in components where no mechanism exists to provide an effective lubricant film. These include: pistons and cylinders, flat-face and spool valves, and sliding vanes in pumps. It is, therefore, common practice to add up to 0.5 percent or 1 percent of a

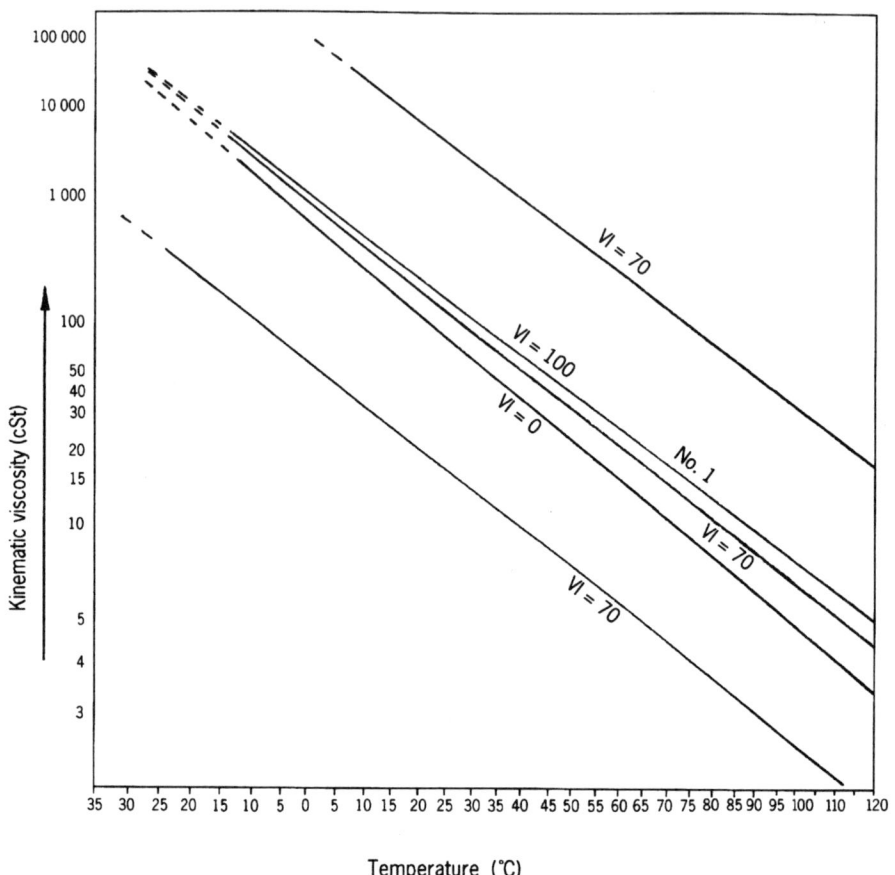

Figure 4.4   Variation of mineral oil viscosity with temperature

phosphorus-containing anti-wear additive, such as tri-aryl phosphate or ZDDP, to hydraulic fluids.

Mild wear can also easily occur in most types of gear and in rolling bearings with cages. Many simple mechanisms, however, like hinges, pivots, and latches are likely to operate entirely under boundary conditions.

For the user, the presence of mild rubbing is usually easy to detect when components are dismantled. It shows initially as a polishing of a surface, and disappearance of the original machining marks. Where it is more fully developed, there will often be an obvious change in dimensions or shape, but without any roughening of the surfaces, or symptoms of over-heating, tearing, gouging, or seizure.

If mild wear seems to be taking place at a faster rate than normal, then it may be that a component is not properly aligned, or is experiencing frequent transient overloads, or even partial lubricant starvation. An attempt should always be made to analyse the cause and correct it.

However, anti-wear additives are mild chemicals which are unlikely to cause any damage to materials or components. This means that a change in lubricant to one having an anti-wear additive, or a more powerful anti-wear additive, may help to reduce a mild wear problem, and is never likely to cause any harm.

The use of extreme-pressure (EP) additives is more complicated. It is also likely to be more critical, because the damage they are intended to combat is much more serious, and because they can produce undesirable side-effects. There are several specific guidelines to the situations in which EP additives should be used, but these are much less precise than are guidelines to viscosity selection.

In ball or roller bearings the need for EP additives depends on the dynamic load on the bearings. Manufacturers' catalogues quote a dynamic load rating C for every type of bearing. They also indicate how to calculate the equivalent dynamic load P, which takes into account the radial and axial loads on the bearing in operation.

It is recommended that EP lubricants should be used in ball bearings when P is greater than 0.25 C, and in roller bearings when P is greater than 0.15C.

It is also recommended that EP lubricants should be used in ball or roller bearings when the lubricant film is less than 20 percent thicker than the combined surface roughness. This may also apply to some types of gear where the contact loads are high, especially where the sliding speed between teeth is high, as in spiral bevel gears. EP oils must always be used in hypoid gears.

In general these rules about the use of EP additives do not take account of bearing speeds. The conventional EP additives containing phosphorus or sulphur depend on high asperity contact temperatures for their action, and these tend to arise at high sliding speeds. Such additives are also fully dissolved in the oil, so that there are no flow problems in high-speed bearings.

Where high contact pressures and scuffing occur in low-speed bearings, solid lubricant additives like molybdenum disulphide or graphite can give huge increases in load-carrying capacity. However, they are not suitable for use in bearings or gears operating at high speed because they have an adverse effect on the flow properties of the oil. In addition, they cannot be used in low-viscosity oils because they tend to sediment out.

The most powerful chlorine and sulphur-containing EP additives are needed in metalworking lubricants. This is a specialized subject which is discussed separately.

Table 4.3 Additives used for specific equipment

Equipment	Additives used
Petrol engines	Anti-oxidant, corrosion inhibitor, viscosity index improver, detergent/dispersant, anti-wear
Diesel engines	Anti-oxidant, corrosion inhibitor, detergent/ dispersant, anti-wear, anti-foam, basic additive (to neutralize acids)
Steam turbines, compressors	Anti-oxidant, corrosion inhibitor, anti-wear, demulsifier
Gears, spur or bevel	Anti-wear, anti-oxidant, anti-foam, sometimes corrosion inhibitor
Gears, spiral bevel, or hypoid	Extreme pressure, anti-oxidant, anti-foam
Gears, worm	Anti-oxidant, corrosion inhibitor, mild boundary additive (e.g., fatty oil)
Machine tool slideways	Friction controller, anti-oxidant, corrosion inhibitor
Hydraulic systems	Anti-oxidant, anti-wear, anti-foam, corrosion inhibitor, pour point depressant, viscosity index improver
Automatic transmissions	Anti-oxidants, anti-wear additives, friction modifiers, corrosion inhibitors

4.4 Other additives

Table 4.3 lists various types of equipment with the additives which are usually used in them. Most of the additives have been mentioned previously.

Anti-foam additives are usually low-viscosity dimethyl silicones or other polymers. They can be useful in any system where the surface of a lubricant is highly agitated. However, in some cases, especially hydraulic systems, careful design of the system can eliminate foaming problems.

Friction modifiers are used to provide the critical frictional properties of automatic transmission fluids, and prevent juddering or squealing when gear changes take place. Similar compounds are also used in limited-slip differentials and lubricated clutches.

The most complex additive packages are used in automotive engine oils, which are discussed separately.

4.5 Selection of base oil

It was stated earlier that well over 90 percent of lubricants are based on mineral oils, because of their availability, low cost, and suitability for the majority of applications. Their wide suitability is, of course, because so much lubricated equipment has been developed in association with mineral oils. However, not all mineral oils are equally suitable for all applications.

Naphthenic oils are particularly suitable for low temperatures, such as in refrigerator oils. They are also better solvents than paraffinic oils and this has two different advantages: (a) the technical problem of dissolving or dispersing many additives is simpler with naphthenic oils; and (b) they are also better at keeping contaminants or degradation products in solution.

The disadvantages of naphthenic oils are mainly at higher temperatures, since their viscosity index and oxidation resistance are both inferior to those of paraffinic oils. There is a continuing trend to higher operating temperatures in all sorts of plant and equipment, and the demand for paraffinic lubricating oils has grown more quickly than the demand for naphthenic lubricants.

This has resulted in higher prices for paraffinic lubricants, so that naphthenic oils tend to have the extra advantage of lower cost. However, this difference is not as marked as it might be. Naphthenic oils are widely used for non-lubricant purposes, such as in fuel oils and in agricultural spray oils and rubber extender oils, where their better solvent properties are beneficial. They are also used as base stocks for hydrocracking processes, to produce high-grade paraffinic base oils.

Conversely, paraffinic oils are preferred where operating temperatures are high, because of their high viscosity index and good oxidation resistance, particularly with anti-oxidants.

The use of non-mineral oils is correspondingly small, and some of the major uses have been mentioned in Chapter 3. Table 4.4 lists the most important properties and some of the principal applications. Synthetic oils are always more expensive than mineral oils, and are used only where their particular advantages justify their cost. The two most frequent reasons for using non-mineral oils are for higher temperatures and for fire resistance.

4.5.1 *High temperatures*

The approximate temperature limits for mineral oils are shown in Fig. 3.5. Those for some common synthetic oils are shown in Fig. 3.6. These give some guidance to the selection of base oils for particular temperatures. However, among the high-temperature oils only synthetic hydrocarbons, polyglycols, di-esters, and polyol esters are good lubricants, and all three are extensively used in high-temperature situations. Synthetic hydrocarbons are becoming widely used in engine oils, but are also used in high-temperature greases. Di-esters and polyol esters are used in very large quantities as aircraft gas turbine engine

Table 4.4 Properties of some synthetic lubricants

Fluid ⟍ Property	Di-ester	Polyol ester	Synthetic hydrocarbon (PAO)	Typical phosphate ester
Maximum temperature in absence of oxygen (°C) (T_D)	270	310	340	120
Maximum temperature in presence of oxygen (°C) (100 hrs)	190	220	150	120
Minimum temperature due to increase in viscosity (°C)	−50	−40	−50	−55
Density (g/ml)	0.91	1.01	0.87	1.12
Viscosity index	145	135	130	0
Boundary lubrication	Good	Good	Fair	Very good
Suitable rubbers	Nitrile, silicone	Nitrile silicone	Nitrile	Butyl, EPR
Effect on plastics	May act as plasticizer	May act as plasticizer	Slight	Powerful solvent
Resistance to attack by water	Good	Good	Excellent	Fair
Effect on metals	Slightly corrosive to non-ferrous metals	Slightly corrosive to non-ferrous metals	None	Enhance corrosion in presence of water
Applications	High temperatures; gas turbines	High temperatures; gas turbines	High temperatures	Fire resistance

Typical methyl silicone	Typical phenyl methyl silicone	Polyglycol (inhibited)	Chlorinated diphenyl	Polyphenyl ether	Typical perfluoro polyether
315	350	260	315	450	380
210	260	200	145	350	300
−50	−30	−20	−10	5	−30
0.97	1.06	1.02	1.42	1.19	1.84
200	175	160	−200 to +25	−60	280
Fair but poor for steel on steel	Fair but poor for steel on steel	Very good	Very good	Fair	Poor
Neoprene, viton	Neoprene, viton	Nitrile	Viton	(None for very high temperatures)	(None for very high temperatures)
Slight, but may leach out plasticizers	Slight, but may leach out plasticizers	Generally mild	Powerful solvent	Polyimides satisfactory	Some softening when hot
Very good	Very good	Good	Excellent	Very good	Very good
Non-corrosive	Non-corrosive	Non-corrosive	Some corrosion of copper alloys	Non-corrosive	Very slight but unsafe with aluminium and magnesium
High temperature; chemical stability	High temperature; chemical stability	Mixed with water; ovens; worm gears	Fire resistance	Very high temperature	Very high temperature

lubricants, and in small quantities in automotive engine oils and metal rolling oils.

Polyglycols are used as gear lubricants, and blended into certain aircraft gas turbine engine oils. One important use is in ovens, or furnaces, where the temperature is so high that any lubricating oil will decompose. Polyglycols have the major advantage that when they decompose they do so cleanly, without leaving any deposits of carbon or ash. For this type of application they are often used with graphite dispersed in them to provide a second lubricant stage after the polyglycol has decomposed.

Silicones, polyphenyl ethers, and perfluoropolyethers all have poor boundary lubrication performance, which cannot at present be improved by additives. As a result they can only be used where contact pressures are low or full lubricant film separation can be maintained.

4.5.2 *Fire resistance*

There are two categories of fire-resistant base oil: the fully synthetic ones and those containing water.

The water-containing oils are obviously limited in use to the temperature range 0 °C to about 80 °C. The oil-in-water emulsions contain only a small proportion of mineral oil (less than 5 percent) to give some lubrication performance, but are still poor lubricants. They are used as hydraulic fluids (HFAE fluids) and in metalworking. The water-in-oil or invert emulsions contain at least 30 percent of water to give effective fire resistance, but they are good lubricants, and are also used as hydraulic fluids (HFB fluids). Finally, the water-polyglycol fluids (HFC fluids) also contain at least 30 percent of water for fire resistance. With up to 60 percent of water they are good lubricants, but they are very widely used with 95 percent water as hydraulic fluids, especially in mining; in those concentrations they are mediocre lubricants.

For fire resistance at higher temperatures three types of synthetic oil can be used. The most common are phosphate esters, which have excellent fire resistance, and are excellent lubricants, but decompose fairly rapidly above about 120 °C. Chlorinated biphenyls are also good lubricants which can be used to about 145 °C. For even higher temperatures perfluoropolyethers have excellent fire resistance and can be used up to about 300 °C, but are very poor lubricants and are extremely expensive.

4.6 Automotive engine oils

Any discussion of lubricating oils would be incomplete without consideration of oils for vehicle engines. These oils are used in greater quantity than all other lubricants combined, and are of interest to far more people than any other lubricants. However, the selection of motor oil for the private owner should almost always follow the recommendations of the vehicle manufacturer. This section is intended mainly for commercial fleet operators, as well as for general interest.

For more than fifty years almost all engine oils have been mineral oils. Prior to that period considerable quantities of vegetable oil were also used, the most important being castor oil. Castor oil is now only used for a few antique cars and aircraft, but where it is used, care must be taken that mineral oil and castor oil are not mixed, as the two types are not compatible.

The oils used for diesel or petrol engines are generally similar in the type of base-stock used and in their viscosity grades. They differ mainly in their additive contents, but even those are becoming increasingly alike.

The viscosity classification used for engine oils is usually the one listed in the SAE table described in the Appendix to Chapter 3 and shown in Table 3A.1. In the past, many oils have been single-grades, i.e., meeting only one of the categories listed in Table 3A.1, such as SAE 20 or SAE 30. Most oils for petrol engines are now multigrades, meeting both a low-temperature requirement and a requirement at 100 °C. Single grade oils are, however, still often used for diesel engines.

Oils for petrol engines are now usually SAE 10W/40 or 15W/40. They have the advantages of providing easy starting at low temperatures with adequate film thickness at operating temperatues. These grades are less viscous than those used twenty years ago, and the lower viscosity has helped to reduce engine friction and improve fuel consumption. The high viscosity index is obtained partly by the use of polymeric viscosity-index improvers, and partly by the use of highly paraffinic base oils.

The additives used in oils for petrol engines also include anti-oxidants, anti-wear additives, corrosion inhibitors, anti-foam additives, detergents, dispersants, and often pour point depressants. Their performance is commonly rated by the API rating scheme, which is described in more detail in Chapter 11. Most modern oils in developed countries are rated SF, SG, or SH.

Oils for diesel engines may be single-grade, but there is an increasing consumption of multigrades such as 15W/30 or 15W/40. Moderate viscosity index may be achieved simply by the use of good-quality base oils, but otherwise a viscosity-index improver may be used.

The other additives used in diesel lubricating oils include those used for petrol engines, although often at higher concentration. They also include additives designed to neutralize acids formed from combustion of sulphur in the fuel, a typical one being an 'over-based calcium additive' in which an excess of lime (calcium carbonate) is kept in dispersion by means of a detergent such as calcium sulphonate.

Diesel engine oil performance is also rated by an API rating scheme described in more detail in Chapter 11. Most modern oils in developed countries are rated CD, CE, or CF, (including CD-II, CF-2, and CF-4).

Developments in petrol engines in recent years have included integral gearboxes and transaxles, supercharging, and fuel injection. These have increased the severity of the lubrication requirements, and some manufacturers now require the oils they recommend to meet a diesel oil rating as well as a petrol-engine rating. As a result, an oil brand may carry a double rating such as CD/SF, indicating that it meets both CD and SF performance standards.

In recent years there has been a gradual increase in the use of 'synthetic' engine oils, especially for petrol engines. These consist mainly of synthetic hydrocarbons. However, synthetic hydrocarbons have a reduced tendency to cause the small amount of swell which is useful for seal rubbers, and may even cause them to shrink. A small amount of a synthetic ester is therefore often blended into a synthetic hydrocarbon in order to give it the same rubber swell characteristics as a typical mineral oil.

Synthetic oils have better resistance to oxidation than a conventional mineral oils, although some specially refined mineral oils are as good as synthetics. These oils can therefore withstand higher operating temperatures. The small amount of added synthetic ester also improves the boundary lubrication.

Two environmental concerns are putting pressure on the development of engine lubricants. The first is to reduce emissions which arise from the oil. This encourages the use of synthetics, which have relatively low volatility for a particular viscosity grade. However, the second is for improved bio-degradability to cope with the problem of the considerable quantities of waste oil which enter the environment. Synthetic hydrocarbons have poor bio-degradability, and much research is under way to find suitable alternative base-stocks.

4.7 Metalworking lubricants

The properties required of metalworking lubricants differ considerably from those required in ordinary bearings and gears. The surfaces which have to be lubricated are the contacting surfaces of the tool and the workpiece. It is often no longer possible to separate the two surfaces, but they must not seize or break, and the tool must not wear out too quickly.

Metals seize, or weld together, much more readily at high temperatures, so one of the most important properties of a metalworking lubricant is not lubrication, but cooling. There is, however, another factor in metalworking which tends to cause seizure.

It was explained in Chapter 1 how freshly exposed metal surfaces, without a protective oxide film, will easily weld together. In metalworking the main result is to create freshly exposed metal surface, so that there is a very high tendency to adhesion.

The actual requirements vary considerably between different metalworking processes. It is impossible to give guidance on the selection of the best coolant or lubricant for a particular process without either going into great detail about composition or listing various manufacturers' brand names. Listing brand names is in any case of limited value because products may be modified fairly often.

This is one area in which the user must generally rely on the literature of the oil suppliers, but most major suppliers give a great deal of guidance. One useful independent listing is given in the *Tribology Handbook* (Neale (editor), 1992, Butterworth Heinemann), but after twenty years, many of the brand names listed have changed.

The following paragraphs are only intended to give a few examples of the types of lubricant needed for different metalworking processes.

(a) *Extrusion, tube drawing, and wire drawing*
These are examples of processes in which very high loads are used to produce very high deformations of the workpiece. The rate of movement may or may not be exceptionally high, but the main problem is to reduce the friction to permit movement to occur at all. The general approaches are to use highly viscous materials or low friction materials. The viscous materials include soaps, chlorinated waxes, and animal fats, or molten glass for high temperatures. The low friction materials include solid lubricants and chlorinated or sulphurized boundary additives, while a vast variety of alkyl and metal stearates are used which probably help to provide viscosity and low friction.

(b) *Rolling*
The object of the lubricant in metal rolling is to control the friction and to prevent the rolled stock from sticking to the rolls. Rolling oils include kerosene, synthetic oils, and many types of emulsion, all including various mild boundary additives such as stearates and oleates.

The formulation and maintenance of rolling emulsions is quite critical. The amount of oil present in the emulsion must be kept above the minimum needed to reduce roll wear and prevent the rolled strip from sticking to the work-rolls. If the emulsion is too stable, or 'tight', then the oil droplets will be small, and will not readily 'break' to release oil to lubricate the strip and rolls. On the other hand, if the emulsion is too unstable, or 'loose', then oil will be released too readily. The friction between roll and strip will then be too low, and the work-rolls will not grip the strip and draw it forward through the roll gap.

(c) *Metalcutting*

In metalcutting operations, grinding, drilling, boring, turning, milling, and so on, a much higher proportion of freshly exposed metal surface is produced than in any of the so-called 'chipless' forming processes, and the surfaces can become very hot. These two factors lead to a very severe tendency for the chips or the workpiece to adhere to the cutting tool. The problems are to cool the tool and to prevent sticking.

The best coolant is water, and most metalcutting lubricants are water-based. At one time mineral or vegetable oils were used, but the use of water has increased through emulsions of 5–10 percent of oil in water, so-called 'soluble oils' with 0.5–1.0 percent of oil in water, to the so-called 'synthetics' in which the necessary additives are dispersed or dissolved in water without any oil. Soluble oils are now often more correctly called 'water mix metal working fluids'. Polyglycol/water solutions are also used as metalcutting lubricants. The emulsifiers used are generally organic rather than metal-salt detergents. Sticking of chips or workpiece to the cutting tool is reduced or eliminated by the use of mild stearate or oleate boundary additives, or powerful EP additives containing sulphur or chlorine.

Efforts have been made by machine tool and lubricant manufacturers to enable the same lubricant to be used throughout a machine tool, as bearing and slideway lubricant, cutting fluid, and hydraulic fluid. As the proportion of oil in the cutting fluid has decreased, it has become increasingly difficult to lubricate bearings and slideways with it. There therefore has been only very limited success in using the same lubricant for all three purposes.

Emulsions are very susceptible to microbiological attack by bacteria and fungi, so it is now normal practice to include a biocide in cutting fluid formulations.

4.8 Process fluids as lubricants

The term 'process fluids' is used here to include any liquids which are present in or near the equipment for some purpose other than lubrication. They may be coolants, refrigerants, fuels, raw materials or products in the chemical industry, syrups or melts in the food industry, treatment liquids or liquors in tanning or paper-making, and so on.

Almost any of these liquids can be used as a lubricant, but they present one problem which makes them quite different from ordinary lubricants: their properties are fixed by the main purpose of their existence, and their composition or properties cannot usually be changed to make them better lubricants. It follows that with process fluids, to a much greater extent than with ordinary lubricants, the bearing must be designed to suit the lubricant.

The variety of problems in using process fluids as lubricants is so great that it may be simplest to illustrate their use by a few examples.

(a) *Refrigerants and cryogenic liquids*
The main problem in using refrigerants or cyrogenic liquids (i.e., liquified gases such as liquid oxygen, nitrogen, or methane) as lubricants is that their viscosities are far too low to give hydrodynamic lubrication. On the other hand, they are such powerful coolants that they can reduce or even completely prevent those types of wear or seizure which are encouraged by frictional heating. They can therefore be used in ball or roller bearings or in polymeric plain bearings, provided that the speed and load are kept low enough to prevent overheating. Pumps in refrigeration systems may be lubricated in this way.

(b) *Syrups or melts*
These liquids, as well as others such as paints and low molecular weight polymers, have high viscosities, and can give very effective hydro-dynamic lubrication. They can be used to lubricate plain bearings or even certain types of gear. One problem is that the viscosity usually changes considerably with temperature, so that heating due to friction or high viscous drag can lower the viscosity and reduce their lubricating ability. Alternatively, if they cool in the bearing they may become too viscous to move. To avoid these problems it may be necessary to surround the bearing with a heating jacket to keep it at a constant temperature, and once again to ensure that the speed and load are not high enough to cause unacceptable frictional heating. Syrups and molasses have been used in this way to lubricate pump bearings in sugar refineries.

(c) *Treatment liquors, brines and chemicals*
These are again often liquids with low viscosities, but without very low temperatures. They can be used in polymeric plain bearings, provided that they do not attack the bearing material. Alternatively, they can be used in 'hydrostatic' or externally pressurized bearings. Described briefly in Chapter 1, these are bearings in which the liquid is pumped in under pressure so that the load is carried by the externally applied pressure.

4.9 Rationalization of lubricating oils

This chapter has dealt with the selection of the best oil for each component. However, by applying these methods to every bearing or gear in a large factory, or even a large machine, the total requirement may be scores, or even hundreds, of different oils.

There are strong arguments for cutting down severely on the number of different oils being used in a plant.

(a)   With a large variety of oils, the purchasing and storage problems are increased.

(b)   If a smaller variety of oils can be used, they can be bought in larger quantities, with the advantage of bulk price reductions.

(c)   With fewer oils in the factory there is less risk of the wrong one being used in a particular application.

The process of reducing the number of different oils used in a factory is called lubricant rationalization, and depends on two considerations: (a) the oil viscosities, and (b) the type of base oil and the additives present.

With regard to viscosity, it was explained earlier that the choice is often not critical, since if a more viscous oil is used the temperature will rise and the viscosity will drop. As long as the increase in viscosity is not too great, the results may well be acceptable.

In practice it has been found that for most applications an increase of 30–50 percent over the ideal viscosity is usually harmless. As a result, national and international standards have been developed in which the whole range of lubricating oils has been reduced to only eighteen different viscosity grades, and many engineering organizations will find that they need only three or four of these grades.

The eighteen grades are listed in Table 4.5, reproduced from the British Standard BS 4231: 1982 'Viscosity Classification for Industrial Liquid Lubricants' which is technically identical with ISO 3448 and other national and international classifications.

Notice that the viscosities are quoted at the reference temperature of 40 °C, and the British Standard includes a chart to convert the values to three other common reference temperatures, 20 °C, 37.8 °C, and 50 °C for three different viscosity index values.

The rationalization procedure is, therefore, to work out the required viscosity at operating temperature and select the viscosity index, then to convert to the equivalent at 40 °C and select the grade giving the nearest *higher* viscosity.

Further rationalization can be achieved by choosing base oil quality and additives such that the performance of the chosen oil meets the most severe requirements for all the various applications. For example, the oxidation resistance should be good enough for the highest temperatures, the boundary lubrication adequate for the most severe loadings, and the pour points low enough for the coldest starts.

Obviously the effect of this will be to specify a higher quality of oil than is needed for some of the applications. The cost of this will need to

Table 4.5   ISO viscosity classification

ISO viscosity grade	Mid-point kinematic viscosity mm^2/s at $40\,°C$	Kinematic viscosity limits mm^2/s at $40\,°C$	
		Minimum	Maximum
ISO VG 2	2.2	1.98	2.42
ISO VG 3	3.2	2.88	3.52
ISO VG 5	4.6	4.14	5.06
ISO VG 7	6.8	6.12	7.48
ISO VG 10	10	9.00	11.00
ISO VG 15	15	13.5	16.5
ISO VG 22	22	19.8	24.2
ISO VG 32	32	28.8	35.2
ISO VG 46	46	41.4	50.6
ISO VG 68	68	61.2	74.8
ISO VG 100	100	90.0	110
ISO VG 150	150	135	165
ISO VG 220	220	198	242
ISO VG 320	320	288	352
ISO VG 460	460	414	506
ISO VG 680	680	612	748
ISO VG 1000	1000	900	1100
ISO VG 1500	1500	1350	1650
ISO VG 2200	2200	1980	2420
ISO VG 3200	3200	2880	3520

be offset against the benefits of the rationalization. Even more important, every chosen oil should be considered against the specific applications, since problems can occasionally be caused by using additives in the wrong place.

4.10   Summary

The guidance given in this chapter for selecting lubricating oils may seem complicated. It may help to summarize the different factors as a sort of checklist to be used whenever a lubricating oil is chosen.

(a)   Work out the best viscosity for each bearing or gear at the expected operating temperature, either by calculation or by using Table 4.1, Table 4.2, or Figs 4.1–4.3.

(b)   Decide what viscosity index is needed in order to cover the whole temperature range from coldest starting temperature to hottest operating temperature. (Or because you have to use a high VI oil to obtain the necessary oil quality.)

(c) Convert the viscosity at the operating temperature to the viscosity at the reference temperature, using Fig. 3.2, Fig. 4.4, or ISO 3448 (BS 4231).

(d) Decide which ISO, BS 4231, or SAE grade meets the needs.

(e) Decide what additives are needed.

(f) Decide if a mineral oil will meet the needs, or whether a natural or synthetic oil is needed.

(g) If one machine contains several different lubricated components, decide whether the same oil can be used for a number of them, and thus use a single lubrication system.

Chapter 5

Oil Supply Systems

5.1 Advantages of oil supply systems

Oil lubrication can be made much more effective if some type of re-supply or feed system is used to supply fresh oil to the lubricated surfaces. The benefits obtained can be any of the following, depending on the type of feed system used.

(a) *Replacement of used oil by fresh oil*
In this way the oil change period is extended because oxidized or otherwise degraded oil is removed, and replaced by clean oil with a fresh supply of additives.

(b) *Removal of contaminants*
Harmful wear debris and other contaminants are removed from the bearing surfaces, and in some cases can then be filtered or centrifuged from the oil for re-circulation to the bearings.

(c) *Cooling*
The temperature of the bearings can be reduced by removal of hot oil and replacement with cooled oil.

There is a very wide variety of different oil feed systems available. They differ in the degree of complication, and, therefore, in cost and reliability, and in the extent to which they achieve the three benefits listed above. For simplicity they can be grouped and described under the following four broad types:

- Total loss (may or may not be centralized)
- Ring, disc, and splash (self-contained)
- Mist and fog (centralized)
- Circulation (centralized)

Guidance on the layout of centralized systems is given in British Standard BS 5807 'Recommendations for centralized lubrication as applied to plant and machinery'.

5.2 Total loss systems

All total loss oil feed systems are designed to supply a small quantity of oil to each bearing or other lubricated component, without any attempt to control or collect the old oil which is being replaced, or any excess of fresh oil which is supplied. It follows that total loss systems are simpler because they do not need return lines or any of the components which are otherwise used to control and clean the recycled oil.

On the other hand, because the flow rates are kept to a minimum there is no coolant effect, and there may be insufficient flow to remove wear debris or other contaminants. Total loss is also wasteful of oil, and leads to a general oiliness of the equipment and surrounding floors unless care is taken.

5.2.1 Manual supply

The simplest form of total loss system is an ordinary hand-held oil can. It has great advantages in permitting individual control of the amount of oil fed to each bearing, and with care it can be highly reliable. It may also be preferred where the bearings are far apart, so that long pipe runs would be needed in a centralized system, or where the weight of a centralized system would be unacceptable, such as in an aircraft.

Its disadvantages are almost entirely those of human weaknesses. Unless the operator is very careful it is easy to miss some of the oiling points, or to fail to check that oil is feeding when the pump is operated. It cannot be used where the feed point is too hot to approach, or out of arm's reach, and the use of manual labour may not be economical.

5.2.2 Drip feed

Almost as simple is a drip feed from a reservoir mounted above the bearings. This reduces the labour involved because the reservoir will require less frequent filling, and it may be possible to use it to supply points which are inaccessible to an oil can.

It is more difficult to judge and control the rate of oil flow with a drip feed and as a result there is a tendency to use too much oil. It is also more difficult to detect a blockage in the feed pipe until it becomes obvious that the level in the reservoir is not falling. For this reason it is useful to fit a sight glass to the reservoir and a glass or transparent plastic section in the feed pipe, so that both the oil level and the flow can be seen. Figure 5.1 shows a suitable arrangement for a drip feed.

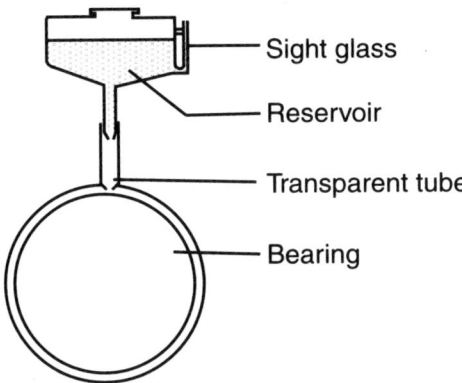

Sight glass

Reservoir

Transparent tube

Bearing

Figure 5.1   Drip feed

5.2.3   *Wick and pad lubrication*

Lubrication by wicks and pads is confined almost entirely to small or lightly loaded plain bearings. The term 'wick' is used to describe a long narrow version, while a pad is of virtually any other shape. In fact they both work in the same way, being wet with oil which they transfer by direct contact to the rotating shaft of the bearing, or in the special case of porous metal bearings, to the outside of the porous bush.

In many cases they may contain only a small initial fill of oil, such as in the combined journal and thrust bearings of small electric motors. They increase the total volume of oil available for the bearing, but ultimately the oil content decreases and the bearing will become starved unless the pad can be refilled, for example from an oil can. In other applications they are used in conjunction with an oil reservoir, and transfer oil from the reservoir to the bearing by a capillary or 'wicking' action. Figure 5.2 shows three possible arrangements.

The oil flow rate is limited, so they cannot be used for large or high-speed bearings. They are useful in dirty conditions because the wick filters dirt out of the oil.

Wick and pad lubrication are very cheap and widely used in low-duty applications.

5.2.4   *Centralized total loss systems*

All the more complicated centralized total loss systems can be looked on as developments of the oil-can or drip feed to provide better control of the oil supply and the cleanliness of the oil, as well as reducing very considerably the labour involved.

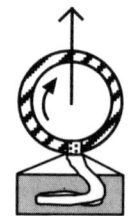
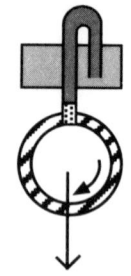

(a) Simple wick   (b) Lower wick   (c) Siphon wick
    without reservoir   with reservoir   with reservoir
    but with facility   for shaft loaded   for shaft loaded
    for adding oil   upward   downward

Figure 5.2   Different forms of wick

The following factors all favour the use of centralized supply systems:

– a large number of bearings using the same oil;
– bearings located close together, so that pipes are short;
– environment unsuitable for staff to enter;
– oil life short, so that frequent oiling is needed.

Their disadvantages are those of complexity, with high initial and maintenance costs. Cleanliness is particularly important when overhauling them, because the fine orifices are easily blocked by particles of dirt.

It is sometimes claimed that centralized automatic systems are more reliable than the oil-can or drip feed, but this is not necessarily so. In all three types the reliability depends on the reliability of the operators, but the greater complexity of centralized systems creates a greater risk of mechanical failure. The results of failing to keep the central reservoir filled in a centralized system will probably be far more serious than failing to oil one bearing with an oil-can. The main advantage of a centralized system from the aspect of reliability is that it is more easily monitored or supervised by more highly trained staff.

Centralized total loss systems are usually manufactured and installed by specialist companies, so it is not proposed to describe them in detail. It may be useful, however, to mention the broad types, the various essential components and some of the factors which need to be considered in selecting them.

Direct systems are those in which the oil passes directly from the pumping device to the lubrication point (bearing, gland, etc.) The pump is usually a positive displacement piston type. The problem of supplying a number of lubrication points can be handled in one of two ways.

A single piston pump with a number of separate outlet ports is a relatively simple economical device where the number of lubricating points is small.

More commonly, multi-piston pumps are used in which a separate piston supplies each outlet. The series of pistons will be operated by a cam or cam-shaft, and the volume of oil delivered to each outlet is controlled by adjusting the stroke of the piston. Each piston is spring-loaded, and the return stroke under the action of the spring recharges each cylinder with oil. Figure 5.3 shows a part of a typical multi-outlet direct system.

Indirect systems are those in which the pump supplies pressurized oil to a complex system or manifold, and the supply to individual lubrication points is controlled by separate metering valves. One of the simplest forms is shown in Fig. 5.4.

In this type of system the pump supplies oil for a brief period at regular intervals. The quantity reaching each lubrication point is determined by the orifice restriction in each valve. A similar type uses a pump to keep the system continuously under pressure, and the meters have very fine capillary orifices.

Other indirect systems use continuous pumps, but the metering valves are operated intermittently.

The following are some of the factors which should be considered in selecting or designing a total loss system.

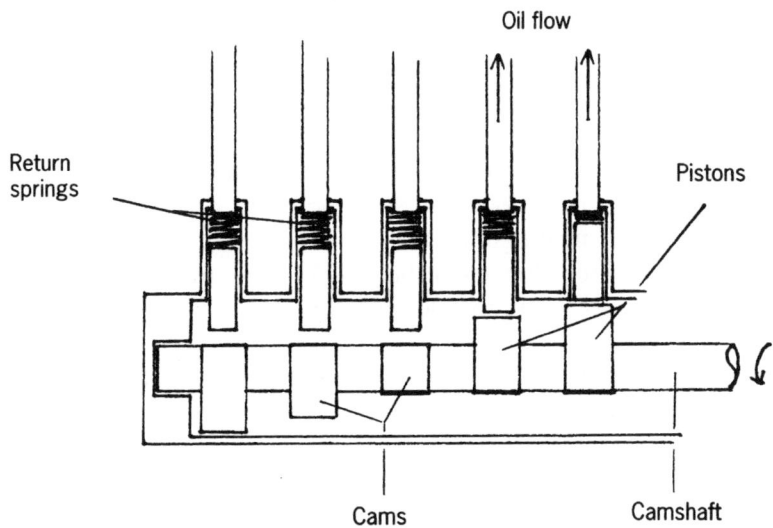

Figure 5.3 Simple multi-outlet direct total loss system

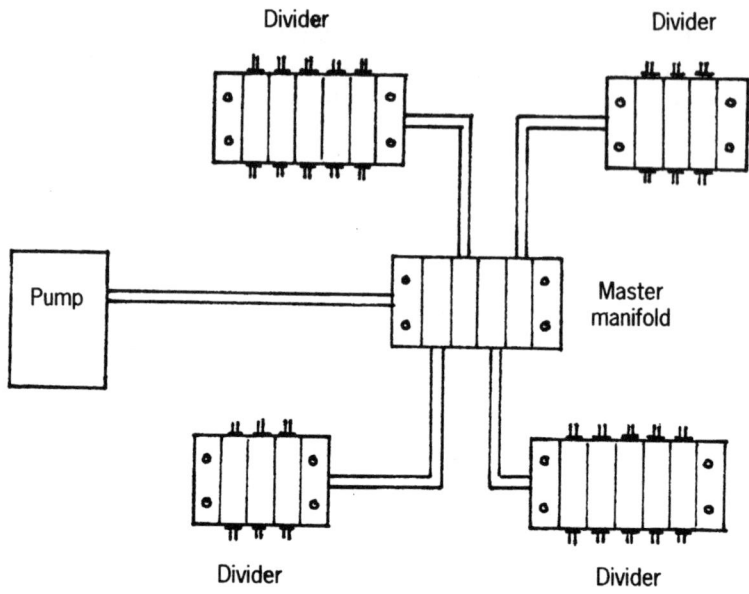

Figure 5.4   Single-line progressive indirect total loss system

(a) *Direct or indirect system*
Direct systems have an advantage in that each lubricating point is
supplied directly by its own piston pump, so that there is greater
certainty that oil will reach the bearing. However, this becomes too
complicated if a very large number of points (more than 100) have to be
lubricated. Indirect systems are mainly used where there are hundreds of
lubrication points.

(b) *Flow rate*
It is not possible to give precise guidance to the required flow rate,
because in practice this depends so much on bearing temperature,
contamination, slight misalignment, and so on.
    The amount required for each bearing varies from about 0.01 to
5 cc/hr (one pint in six years to one gallon in a month depending on the
size and speed of the bearing).
    The guidelines shown in Table 5.1 are published by Interlube Systems
Limited for calculating the amount of oil needed for different
components in ml per hour, and similar systems are used by other
suppliers. The figures apply if the oil viscosity at the operating
temperature is between 50 and 500 centistokes. They assume that the
components are being used within their proper load and speed range,
and are in good condition. If, for example, the clearance in a plain

Table 5.1 Oil supply rate in ml/hr for total loss systems (Courtesy of Interlube Systems Limited)

Rolling bearings	0.003 × Shaft diameter (mm) × No. of rows
	0.08 × Shaft diameter (in) × No. of rows
Plain bearings	0.0002 × Shaft diameter (mm) × Width (mm)
	0.13 × Shaft diameter (in) × Width (in)
Plain horizontal slides	0.00005 × Travel (mm) × Width (mm)
	0.032 × Travel (in) × Width (in)
Plain vertical slides	0.0001 × Travel (mm) × Width (mm)
	0.064 × Travel (in) × Width (in)
Anti-friction bearing ways	0.0006 × Travel (mm) × No. of rows
	0.015 × Travel (in) × No. of rows
Cams	0.0003 × Maximum diameter (mm) × Width (mm)
	0.2 × Maximum diameter (in) × Width (in)
Gears	0.0004 × Largest gear diameter (mm) × Width (mm)
	0.25 × Largest gear diameter (in) × Width (in)
Chains	0.00008 × Length (mm) × Width (mm)
	0.05 × Length (in) × Width (in)

bearing is unusually high, perhaps due to wear, then the rate of oil loss from the bearing will be higher, and a correspondingly high feed rate will be needed.

The table is very simple to use. For example, a plain bearing 30 mm in diameter by 20 mm long requiring a 70 cSt oil at operating temperature will need an oil feed of $30 \times 20 \times 0.0002 = 0.12$ ml/hr.

However, this should only be considered as a rough initial guide, and a careful eye should be kept on any new installation to make sure that too little oil, or more probably too much, is not being supplied. For critical applications, more accurate, and much more complicated, calculations are available.

(c) *Monitoring supply*

It is important to ensure that oil is actually being fed to each point, especially with indirect systems. This is usually done by positioning sight glasses as close as possible to the lubrication point. In addition, it is essential to make sure that the oil reservoir does not become empty. This can be done by regular inspection, or by a sight glass, or some other type of tell-tale mechanism.

On the whole, total loss oil systems are quite reliable and economical to fit and are very widely used. However, they are generally wasteful of oil, and may become less popular if oil prices increase.

5.3   Ring, disc and splash lubrication

Earlier chapters mentioned the use of a small amount of oil in place in the bearing, and the limits on using this technique because of overheating, contamination, and oil loss or breakdown. It is possible, of course, just to increase the amount of oil present, without making any other changes in the system.

The first problem which then arises is that, as the amount of oil is increased, it starts to leak out of the bearing area. This may be just messy, as in heavy engineering, or absolutely unacceptable, as in food or textile plant. In either case it is wasteful and destroys the benefit of using more oil.

The solution is to seal the various openings, including those through which rotating shafts run. Once this has been done, the whole enclosed space could in theory be filled with oil. There are several reasons for not doing so.

(a)   The main reason is that if the moving parts are completely immersed, the oil becomes continuously stirred. A great deal of power can be lost in this churning of the oil, thus reducing the efficiency of the equipment and causing unnecessary heating of the oil.

(b)   If the oil completely fills the enclosed space then as it gets hot in operation it will need to expand, and high pressures and leakage will develop.

(c)   The shaft seals are less effective when flooded than they are when simply splashed with oil, so that leakage can become even more of a problem.

A compromise is therefore needed. In practice the total amount of oil is increased by expanding the enclosed space to create a larger reservoir, but the level is arranged so that the moving components are only partly submerged. The submerged parts then pick up oil and fling it, giving splash lubrication of the parts which are above the oil level. The extent to which they are submerged depends on the nature of the moving parts. There are some simple guide-lines which can be used.

Effect of speed
Obviously the higher the speed the more effective is the splashing, and the greater the churning losses, so that the depth of immersion should be less in high-speed systems.

Spur and helical gears
The gears normally dip into the oil to about twice the tooth height.

Worm gears
Worm half immersed if worm is below wheel. Wheel immersed between tooth depth and half diameter, depending on speed, if worm is above wheel.

Crankshafts
Crankshafts are nowadays usually supplied with pumped oil under pressure, but in small systems they may be splash lubricated. The oil level is generally arranged to cover the big-end bearings at their lowest position, but in very slow systems it may be high enough to submerge the lowest parts of the main bearings. A scoop-shaped collector is sometimes fitted to the big-end bearings to pick up oil and force it into the bearings.

Chains
Oil level high enough to cover the lowest chain links and sprocket teeth.

Plain bearings – ring or disc lubrication
It is not easy to feed oil into a plain bearing by partly submerging it in the oil, because the bottom part of the bearing usually has the highest pressure and the smallest clearance. Oil must be fed into the low-pressure region higher up in the bearing.

This is commonly done by rotating discs or rings which dip into the oil and transfer it to the top of the bearing. A disc is mounted on the rotating shaft alongside the bearing. It has a hollowed face on the side facing the bearing which picks up oil and throws it into an opening at the top of the bearing, as shown in Fig. 5.5(a). If the speed of rotation is too high the oil will be thrown outwards and will not reach the bearings,

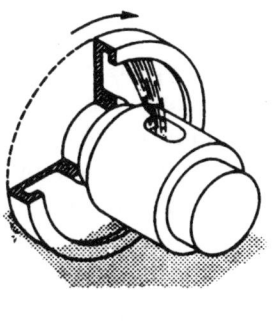

(a) Disc

(b) Ring

Figure 5.5   Ring and disc lubrication

but disc lubrication can be used if

Shaft diameter (mm) × rpm is less than 200 000

Shaft diameter (in) × rpm is less than 8000

or to higher speeds if the oil is cooled.

Rings feed oil directly into gaps in the top of the bearing, as shown in Fig. 5.5(b). They rest on top of the shaft, and as the shaft rotates it causes the rings to rotate, but at a lower speed. Rings are generally used at lower shaft speeds than discs, and with heavier oils, but they can be used if

Shaft diameter (mm) × rpm is less than 150 000

Shaft diameter (in) × rpm is less than 6000

5.4 Oil mist or fog systems

A very versatile technique is to circulate oil in the form of minute droplets dispersed in a stream of air. This is also a total loss system, but differs considerably from those described above.

Compressed air, usually from the general shop supply, is filtered and fed through a *generator*, where it picks up the oil in the form of very fine droplets. The air stream is kept at low speed (6 m/s or 20 ft/sec) and low pressure (less than 50 mbar or 0.75 psig). Under these conditions the oil remains dispersed in the air stream and does not settle out on the walls of the tubes.

The mist is passed through plastic, steel, or copper tubes to the lubrication points, where it passes through fine nozzles or *reclassifiers* which raise the speed to 45 m/s (150 ft/sec) or more and direct it on to the component which is to be lubricated. The tubes must be smooth and clean, without sharp bends, otherwise the oil may settle out. The pipe and nozzle diameters are chosen to divide the oil supply between the various lubrication points.

Once the air speed has been increased, the oil droplets stick to any surface which they strike. The released oil then lubricates the appropriate components while the air escapes to atmosphere.

Methods are available to calculate the oil and air flow needed, but they are not highly accurate. There is again a tendency to use too much oil to give a safety factor, but oil mist systems are generally less wasteful, cleaner, and more reliable than pumped total loss systems.

Oil mists have one other major advantage in that they give very effective cooling, and they tend to be used for high-speed rolling bearings or for chains in hot environments. They are less easy to use with plain bearings because of the difficulty of feeding the reclassified oil from the nozzle into the bearing clearance.

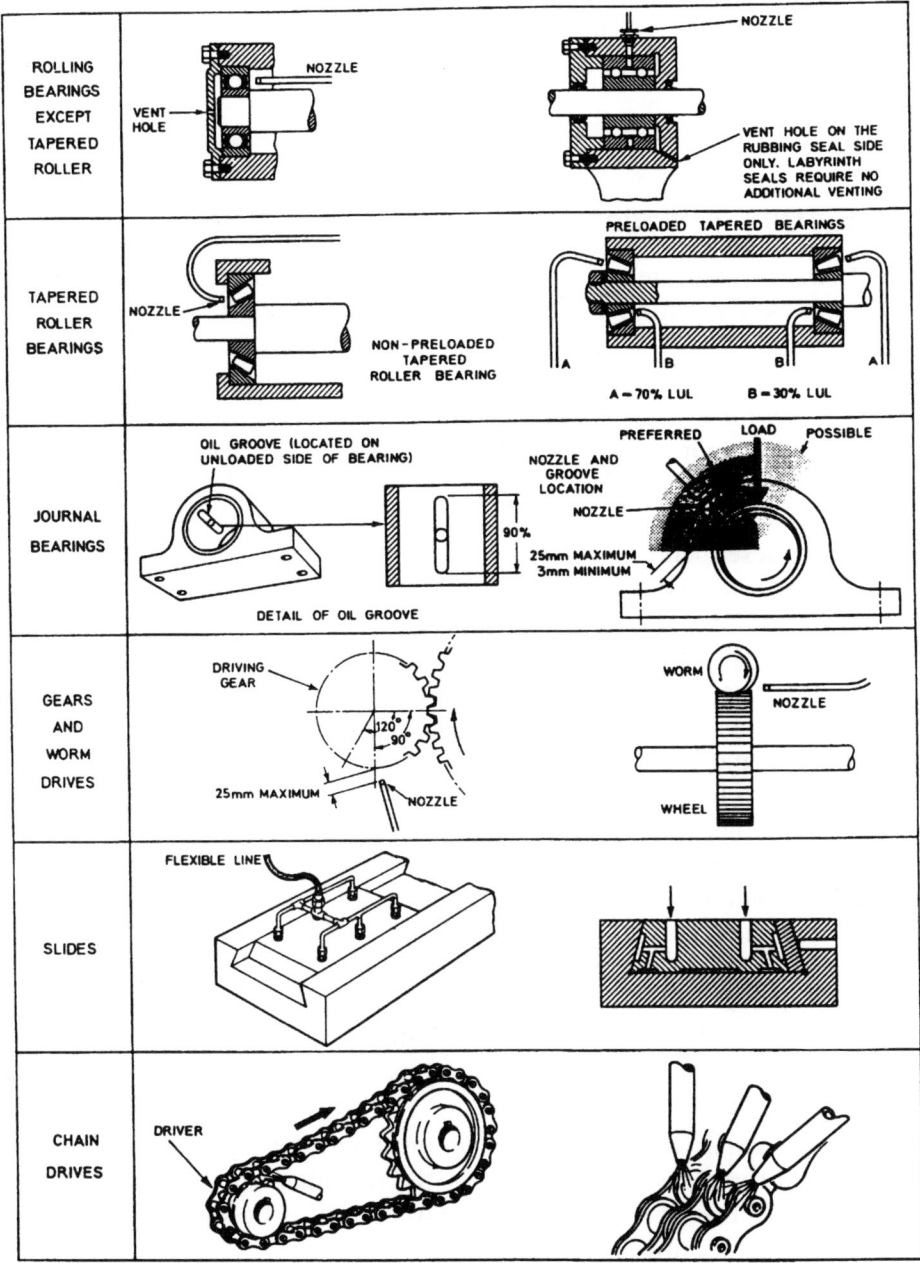

Figure 5.6   Recommended nozzle positions in oil-mist lubrication

One major concern with oil mist systems is the amount of oil which can escape with the air into the factory atmosphere. It can be controlled by choosing the right nozzle diameter, properly positioning the nozzle(s), and especially by choosing a suitable oil. Figure 5.6 shows recommended nozzle positions for a variety of different lubricated components.

5.5 Oil circulation systems

Oil circulation systems differ from all those previously described in that the oil is fed into the bearings, gears, etc., and then returned to a separate reservoir for re-use. This makes it possible to feed much larger quantities of oil without waste.

A basic oil circulation system consists of a reservoir, a pump, a feed line, one or more lubrication points, and a return line.

Where the equipment to be lubricated is small or slow the oil circulation system may be basically similar to the centralized total loss systems described earlier. The only addition needed is the equipment to collect the oil and return it to the reservoir.

For very large systems, or where the speeds are high, and especially where the oil must remove a great deal of heat from the bearings, a different type of circulation system is needed. The oil flow is higher to give both lubrication and cooling, and the circulation system is probably larger and more complicated.

The same basic components are needed but other items may include heaters, coolers, pressure gauges, pressure relief valves, flow dividers, strainers, filters, water separators, chip detectors, sampling points, and standby pumps. These may all be mounted in or on the reservoir in a self-contained system. With very large systems, such as in steel or chemical works, they may be laid out separately to provide easier access for servicing.

A typical self-contained system is shown in Fig. 5.7.

Oil reservoir
The total capacity of the oil reservoir should be equivalent to between 20 and 40 minutes supply of oil at the maximum flow rate. There may be one or more tanks, depending on whether it is desirable to change the oil without shutting down the system.

The reservoir normally has from one to three baffle plates to separate the return points from the pump suction point, and to prevent foam, water or sediment from reaching the pump and being re-circulated. It also has a sight gauge or some other type of level indicator and an air breather.

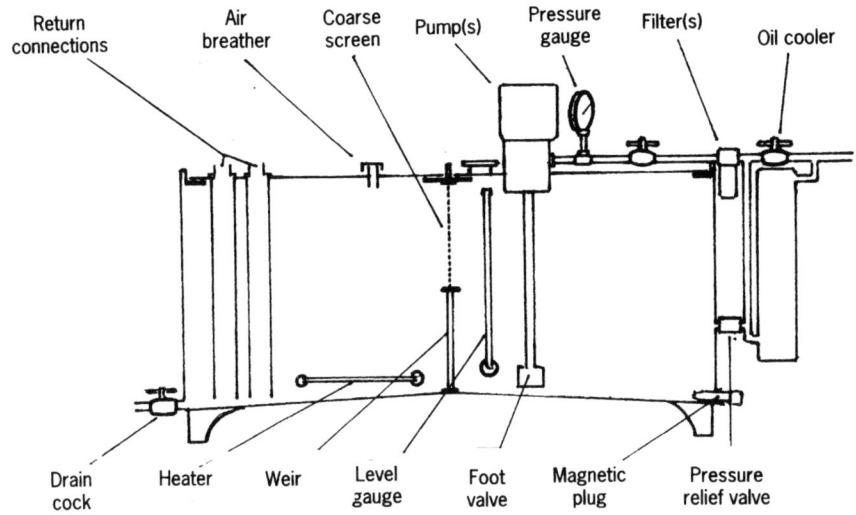

Figure 5.7   Self-contained oil circulation system

Pump

The pump may be of almost any type, including gear, vane, swash-plate, screw or centrifugal, depending on oil viscosity, and flow rate. The total pump capacity should be at least 20 percent higher than the maximum oil flow required, and in most cases it is desirable to provide a standby pump. The inlet to each pump should be fitted with a filter or strainer to prevent solid particles from damaging the pump.

Heaters and coolers

The cooling which takes place in the return lines and the oil tank will often be adequate, but an additional cooling system may be needed and usually it will be water-cooled.

If the oil is very viscous it may be necessary to heat it to reduce the viscosity for easy flow and to improve the separation of contaminants. The heating may be by hot-water coils, steam coils, or electricity. If electricity or high-temperature steam (over 110 °C) is used, it is important to make sure that oil does not break down on the surface of the heater, since the deposits formed will interfere with heat transfer. Figure 5.8 shows how the required heating surface area is related to heat input and flow rate.

It is not unusual for both heaters and coolers to be present in the same system, and in large or critical systems they should be thermostatically controlled.

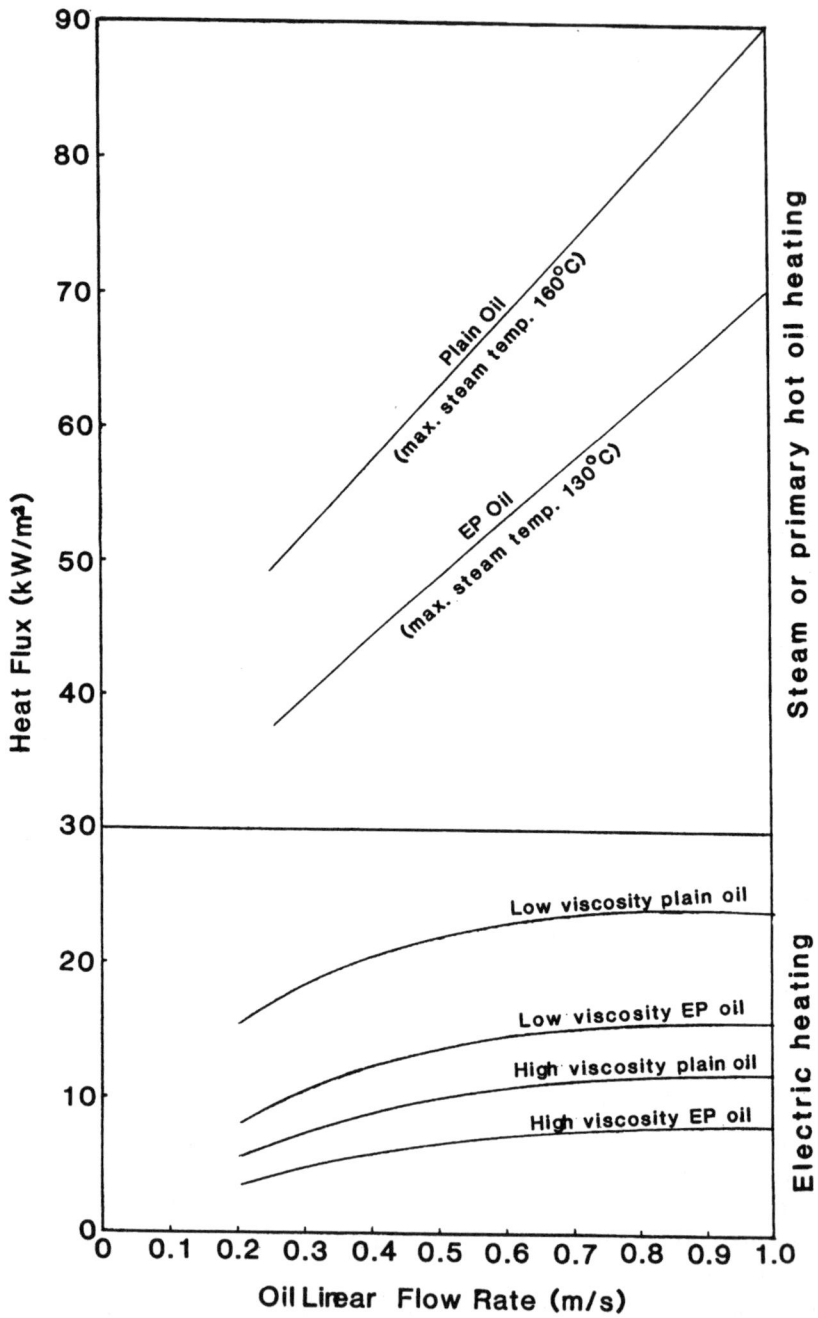

Figure 5.8   Heat flux recommendations for oil heaters

Flow dividers

These are used in the same way as for total loss systems to direct the required amount of oil to each lubrication point. If possible, a sight gauge should be fitted to each outlet to make sure that oil is feeding.

Strainers and filters

Strainers are fitted to the return lines, the pump inlets, and sometimes between the halves of the reservoir to remove large debris particles which might damage components. A magnetic strainer may be used, especially with lubrication of large gears, to collect metallic particles. Some solid material will settle out in the tank if the flow is not too rapid, but filters should also be used to remove the remainder of the solid contaminants down to an acceptable size level. In particularly dirty systems some type of high-capacity self-cleaning or automatically changing screen is desirable.

Water separators

Some water contamination will settle out in the reservoir, evaporate from a hot reservoir, or be removed by the filters. If there is very much water contamination a separate trap may be required, and in extreme cases a centrifugal water separator may be necessary. A centrifugal cleaner which will remove water and sludge is shown in the next chapter.

Chip detectors and sampling points

The use of some form of oil condition monitoring is becoming common for large or important oil systems. This subject is discussed in more detail in Chapter 12, but thought should be given to the provision of detectors and sampling points when the circulation system is first designed.

5.6 Oil changing problems

When designing and installing oil systems it is important to remember that at some time the oil in a circulation or splash system will need to be changed.

One or more drain cocks or plugs will therefore have to be fitted, and the following points should be considered.

(a) *Complete draining*

It is important to be able to drain as much as possible of the old oil out of the system. Any which cannot be drained will either have to be flushed out (an extra expense) or will remain to contaminate the fresh oil. A drain should be fitted at the lowest point in the system, and all pipes and other components should be laid out so that the oil can drain back to that point. If this cannot be done then additional drains should be fitted at every low point which could trap oil, water, sludge, or sediment.

(b) *Access*

The stopcock or plug must be readily accessible for operation, and should be clearly visible. A drain which has not been properly closed is a common cause of accidental oil loss. There must be sufficient clearance for a collecting vessel or hose to be placed beneath the drain.

(c) *Type and size*

Stopcocks are more convenient and much easier to control than plugs, but they are more likely to be blocked by sludge or sediment. They should preferably be large-bore and be installed vertically and not horizontally.

The subject of oil changing is discussed in more detail in the next chapter.

5.7 Selection of the appropriate system

The advantages and disadvantages of the various oil feed systems are summarized in Table 5.2 and the following sections indicate the conditions in which each system is suitable.

(a) *Oil can*

Bearings readily accessible; few in number or some distance apart; no cooling required.

(b) *Drip feed*

Bearings less readily accessible; few in number or far apart; no cooling required; personnel expensive or often not present (unmanned plant).

(c) *Centralized total loss*

Large numbers of bearings close together; no cooling or contaminant removal needed; possible oil contamination of the surrounding area acceptable.

(d) *Oil mist or fog*

Small numbers of bearings fairly close together; cooling needed but not contaminant removal; possible oil contamination of the surrounding area undesirable.

(e) *Wick or pad*

Small, lightly loaded plain bearings; no cooling or contaminant removal needed.

(f) *Ring, disc, or splash*

Individual, isolated systems; cooling and contaminant removal needed.

Table 5.2 Advantages and disadvantages of different oil feed systems

System	Advantages	Disadvantages
Oil can	– Low initial cost – Simple – Easy to check	– High labour cost – Reliability depends on user – No cooling – Easy to use wrong oil – requires close approach to machinery – No recovery of used oil
Drip feed	– Low initial cost – Low labour cost	– Requires close approach to machinery – Care needed to ensure reliable flow – No cooling – No recovery of used oil
Centralized total loss	– Low labour cost – High reliability – Reduced risk of wrong oil	– High initial cost – Higher maintenance cost – No cooling – No recovery of used oil
Oil mist/fog	– Low labour cost – High reliability – Low consumption of oil – Useful cooling	– High initial cost – Careful flow control needed – No recovery of used oil
Wick or pad	– Low initial cost – Low labour cost	– No cooling – May become blocked
Ring or disc	– Low labour cost – Generally reliable – Some cooling – Recovery of used oil	– Fairly high initial cost – Limited speed range
Splash	– Low initial cost – Low labour cost – Recovery of used oil	– Limited speed range
Circulation	– Wide range of use – Generally reliable – Effective cooling – Recovery of used oil – Removal of contaminants	– High initial cost – High maintenance cost

(g) *Circulation system*

A number of components needing lubrication; considerable cooling or very effective contaminant removal needed.

Chapter 6

Oil Changing and Oil Conservation

6.1 Oil changing

Lubricating oil deteriorates in use in several different ways, and with this deterioration comes a risk of damage to the lubricated system. In order to eliminate, or at least reduce, this risk it is eventually necessary to replace the oil; this is, in fact, the most common reason for oil changing.

Another reason for changing oil is to maintain the right viscosity when the operating temperature changes. In temperate climates this is rarely necessary, but at one time it was considered good practice to change the oil in a car engine, using an SAE 30 or even SAE 40 grade in summer and an SAE 10W or 20W in winter. The general use of multigrade oils has made this much less common, but in extreme climates many people will change from an SAE 20/50 grade in summer to an SAE 10W/40 in winter. In extreme climates it may also be necessary to change industrial oils to a less viscous grade in winter.

Yet another reason for oil changing is to introduce a new type of oil where the previous one has proved unsatisfactory in some way. This is probably far less common than changing to a fresh supply of the same oil, but it introduces a few special problems and will be considered separately later.

The effects of oil deterioration and contamination will also be described in detail later. Briefly, the most important effects are:

(a) oxidation, producing acidity which can lead to corrosion;
(b) oxidation, shear, evaporation, or contamination giving change in viscosity which can lead to flow problems or failure to lubricate;
(c) solid contaminants or breakdown products which can cause

abrasion of bearings or gears, blocking of filters or fine passages, or a major increase in oil viscosity;

(d)   contamination with water, which can cause corrosion, interfere with the effects of the additives, or emulsify with the oil causing flow problems or failure to lubricate.

A major difficulty is to decide when to change the oil. Obviously it would be ideal if it could be changed just before any attack on the system or any decline in the quality of lubrication takes place. Unfortunately, the speed at which deterioration takes place can vary considerably, even between systems which are supposed to be identical. Consequently, guidance to the right oil change period is not very easy or accurate.

If we consider again the example of a private car engine, the car manufacturer's recommendation may be to change the oil every 5000 or 10 000 miles, but the rate at which the oil deteriorates will depend very much on the way in which the car is used. A car which is used only for short journeys with several hours between them may never warm up to the proper running temperature, and water and unburned fuel will tend to accumulate in the oil so that it should be changed more often.

On the other hand, an identical car which is used only for long journeys, is well serviced, and driven steadily at a reasonable cruising speed might never need an oil change at all. Replacement of the normal oil consumption by topping up might be enough to maintain the oil quality at a satisfactory level.

In the past hundred years mineral oils have been so plentiful and cheap that it has been normal to change the oil far sooner than was really necessary. If experience showed that in a particular type of system damage due to poor oil condition started after between 200 hours and 1000 hours, then the oil change period would be established as 100 hours. This would provide a safety factor against expensive damage to the system, and the cost of replacement oil would be far lower than the possible cost of not replacing it.

This situation is now changing, and is likely to change more and more rapidly, but at the present time the cost of lubricating oil is still one of the minor costs in servicing plant or equipment. On average the total cost of the lubricating oil used in a vehicle engine throughout its life is likely to be much less than the cost of a replacement engine. The labour cost involved in oil changing may be more significant that the cost of the oil, especially with small oil systems such as in vehicle engines. For many people the need to conserve oil supplies, and the problem of disposing of the used oil without harming the environment may be more important reasons for reducing oil changes.

In a large plant the cost of an oil fill may be £10 000 or more. Even if this is small compared with the value of the plant, or even the cost of servicing the plant, it is high enough for operators to think very seriously about the right oil change period.

The problem is further complicated by the fact that the rate of deterioration is not steady. It was explained in Chapter 3 that oxidation proceeds very slowly until the anti-oxidants are used up, and then takes place much more quickly. It was also mentioned that the presence of certain contaminants leads to more rapid breakdown.

Figure 6.1 shows the way in which the acidity changes in the oil in a diesel engine, and shows clearly the rapid increase in the rate of deterioration as the engine approached failure.

It follows that if the oil change is left a little too late, the result may be serious damage to the system.

One solution is to use various techniques for monitoring the quality of the oil. This subject is discussed in detail in Chapter 12, but for the smallest oil systems it is doubtful if it will ever be economical to monitor the oil. Some guidelines to choosing the oil change period will be given later in this chapter.

6.2 Oil resources and conservation

For over a hundred years mineral oils have dominated lubrication, so it is logical to consider their availability and conservation first.

Mineral oils come almost entirely from petroleum sources. Synthetic hydrocarbon oils are mainly produced from petrochemicals. A small proportion of those may derive from natural gas, but otherwise they also come almost entirely from petroleum.

At various times in the recent past there has been very serious concern about the remaining life of the world's petroleum resources. It was at one time widely accepted that they were likely to be exhausted within human lifetime unless the rate of consumption was reduced. More recently there has been a 15–20 percent reduction in consumption and considerable success in developing new oilfields, so that the concern has diminished. The present industry view seems to be that the rate of oil discovery will continue to keep pace with demand, and that the threat of exhausting the world's supplies has disappeared. This view may in fact be over-optimistic, and there can be no doubt that because petroleum is a non-renewable resource, it must eventually be used up.

Another consideration which makes it desirable to reduce petroleum consumption is the enormous environmental damage caused by its use. This is mainly in a few major transport accidents, but also in thousands of smaller incidents and in continual loss from lubricated machinery and vehicles.

The major use of petroleum is of course for fuel. The pressure for

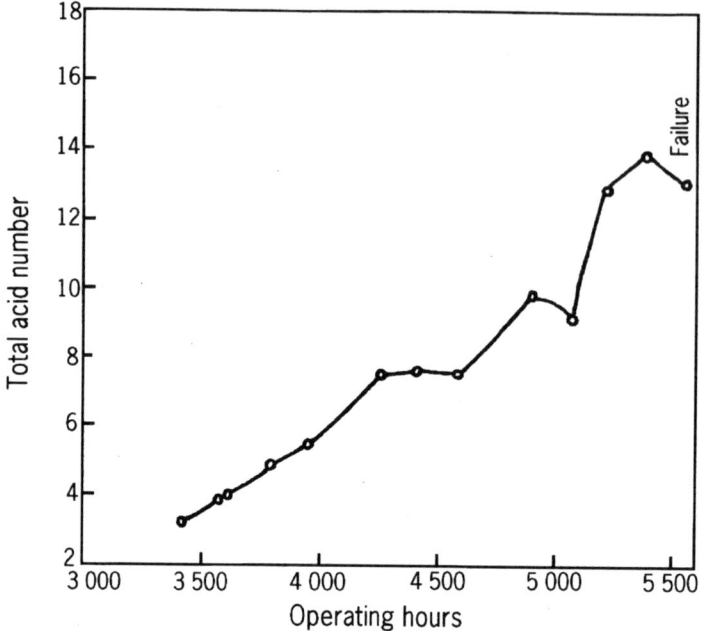

Figure 6.1   Change of acid number in a diesel engine

conservation is therefore mainly seen in terms of fuel consumption, and in the further environmental damage caused by atmospheric carbon dioxide. If all these pressures succeed in reducing the overall consumption of petroleum, then the future supply of mineral oils could be secure for centuries.

It follows that there can be no realistic estimates of the future availability of mineral oils, but it is wise for some thought to be given to alternatives.

The only major alternative source is likely to be coal, which exists in vast quantities. However, a great deal of work remains to be done to produce lubricant-grade mineral oils from coal. Their production will inevitably be more complicated and, therefore, more expensive than the relatively simple process of distilling petroleum.

The second most widely used group of lubricants are the vegetable oils, such as castor oil and rapeseed oil. Their use has declined as the use of mineral oils has increased. There are many different vegetable oils which can be used as lubricants, and there is no doubt that world production could be increased considerably. The oil-producing plants are grown and harvested in the same way as other crops, so that they are a renewable or self-sustaining resource. The ultimate limit on the availability of vegetable oil lubricants is likely to be the competition for suitable land with crops used for other purposes, e.g., food, textiles, etc.

Unfortunately, vegetable oils are less stable than mineral oils, especially at high temperatures, and oxidize to give gums and lacquers. They cannot be used as a direct substitute for mineral oils in most applications, so there would be problems in attempting to use them to fill the gap caused by diminishing mineral oil supplies. However, one pointer to possible future extension of the use of vegetable oils is the recent production in Austria of a diesel-engine oil based on rapeseed, presumably with some degree of chemical treatment to improve the oxidation resistance.

Animal fats and oils are also used to a limited extent for special lubrication problems. They are not, however, likely to be available or suitable for much wider use than at present.

The case of sperm oil, obtained from the sperm whale, is interesting, and indicates the influence of environmental pressure on lubricant use. This oil, in its sulphurized form, has a very remarkable combination of stability with extreme pressure properties. It may be true to say that, as yet, no other additive is really a satisfactory replacement for it. Its availability was rapidly reduced, and it is probably now completely unavailable for two reasons. The first is that killing of whales is no longer acceptable to most civilized people; most countries have banned the use of whale products, including sperm oil. The second is that with the excessive hunting of the larger whales, the remaining whalers turned to hunting smaller ones, including sperm whales, so that their numbers are also likely to be reduced.

The synthetic oils are mainly derived from petroleum, but it would almost certainly eventually be possible to produce them from coal. Although it will be expensive to manufacture them from coal, it is also expensive to make them from petroleum. On the whole, therefore, they may suffer less than mineral oils from any reduced availability of petroleum. They may, in fact, become more competitive in price, and therefore more widely used.

It must be remembered of course that with an increase in the variety of coal-based products the demand for coal will also increase. This will have a further influence in increasing the cost of lubricants.

These probable effects are summarized in Table 6.1, but briefly the

Table 6.1   Probable effects of future lubricant supply problems

1. Decreasing availability of mineral lubricating oils
2. Synthetic oils becoming more competitive
3. Development of coal-based mineral and synthetic oils
4. Increasing use of vegetable oils
5. Increasing oil change periods
6. Much more laundering and re-refining of oils

real cost of lubricants can be expected to increase considerably in the next fifty years. There are also likely to be changes in the relative costs of certain lubricants. As a result it will be important to conserve lubricants by using them more efficiently. This may mean extending oil change periods and cutting down waste. It will also mean more cleaning up and re-refining of lubricants to enable them to be re-used.

6.3   Oil deterioration in use

The factors which bring about deterioration of an oil in use have been described in Chapter 3. Here we will consider the effects on the properties and performance of the oil.

(a)   Oxidation

The most important type of chemical reaction affecting oils in service is oxidation, which very often determines service life. The effect of oxidation is to introduce oxygen atoms into the base oil molecules, converting hydrocarbon molecules into aldehydes and then into acids. These products are less stable than the original hydrocarbon molecules, and so they tend to be further attacked. The final products may be highly oxidized gums and lacquers, or even solid residues, such as the carbonized materials soot and coke.

The rate of oxidation depends on the oxidation resistance of the oil and on the operating temperature. Oxidation is also 'catalysed' by certain metals or, in other words, its speed is increased when those metals are present. This effect is especially strong when the metal surfaces are fresh, so oxidation is greater when iron, steel, aluminium, or copper surfaces are being worn during operation.

The presence of anti-oxidants, either natural or additives, reduces the rate of oxidation very considerably. Eventually, however, the anti-oxidants will be used up, and the rate of oxidation will then increase rapidly. The combustion of fuel in an engine can also produce acidic oxidation products in the oil, especially if the combustion is poorly controlled, or if the fuel contains very much sulphur.

In practice, thermal breakdown in the absence of oxygen is much less common. This is because oxygen is usually present. Its effects, however, may be similar, leading to gums, lacquers, and sludges.

(b)   Breakdown due to shear

When an oil is present between sliding surfaces it is subjected to shear, and large molecules may be torn apart by the mechanical forces involved. This effect is not usually important for base oils, but can have a marked effect on polymeric viscosity index improvers in multigrade oils, or on polymeric synthetics such as silicones or polyglycols. There is also some evidence that the effect is greater at high temperatures.

The result of this shearing of large molecules is a reduction in viscosity; shear breakdown of viscosity index improvers also causes a decrease in the viscosity index.

(c) *Evaporation*

Lubricating oils are not generally volatile. Normally it is only with low-viscosity oils at higher temperatures or in vacuum applications that evaporation is likely to be significant. However, it has been shown that there is a significant loss of the lower fractions of engine oils in the region of the piston ring/cylinder interface. The effect of evaporation is generally to produce an increase in viscosity.

(d) *Contamination with fuel*

The problem of contamination of lubricating oil by fuel is obviously one which occurs only in internal combustion engines or other systems which contain liquid fuels. Similar effects can arise in chemical plant or refineries where liquids having lower viscosity and perhaps higher flammability can enter the oil system.

In engines, the contamination may occur because of poor combustion control, which results in unburned fuel mixing with the oil in the cylinders, especially before the engine becomes fully warm. In petrol-injection or diesel engines it can also occur because of pump or injector leaks. In refineries or chemical plant the most likely source is leakage through a seal.

The main effect of this type of contamination is that it lowers the viscosity of the oil, sometimes to the extent that lubrication fails and severe wear takes place. Another risk is that a fire may be caused because the fuel/oil mixture is more easily ignited, and in fact wear particles or frictional heating can act as ignition sources.

(e) *Contamination by solids*

There are four main types of solid contaminant in an oil system. These are: wear debris, breakdown products from the oil, combustion products (soot) in an engine, and external dust or dirt ingested, or sucked, into an engine or an air compressor. All of these substances can be detected by various oil monitoring techniques and these are discussed in Chapter 12.

Wear debris and outside dust sucked into a system can be very abrasive, and produce severe wear of components like bearings, valves, gears, or cams.

Soot and oil breakdown products can increase the viscosity of the oil to an unacceptable level, and in extreme cases soot has been known to thicken an oil to the consistency of a grease.

Filters are commonly used to remove solid contaminants from an oil system and, provided that the filter is properly designed and selected for the solids present, then the problem will normally be under control. However, if the type of solid material is different from what was expected or in much greater quantity, then the filters may become blocked and this can create additional problems. Any of the types of debris can also block fine oil feeds or holes, and thus cause oil starvation.

The effect of filtration is to remove from the filtered liquid all, or almost all, of the particles larger than some nominal size. If a 5 micron filter is used, then this means that most particles larger than $5\,\mu m$ will be removed. If for simplicity we think of particles being like rectangular blocks, then it is the second-largest dimension which is important (i.e., width rather than length or thickness) and this is called the critical dimension. Particles are never rectangular blocks, but in practice rods or fibres much longer than $5\,\mu m$ will pass through a 5 micron filter, provided their thickness is less than $5\,\mu m$. On the other hand, a flake $6\,\mu m$ long and $5.5\,\mu m$ wide will probably be removed even if its thickness is much less than $1\,\mu m$.

In terms of total numbers of particles, solid contaminants usually contain relatively few of the larger particles, and very large numbers of extremely small particles. Filtration may remove most of the large particles, but have little effect on the finest particles. The finest particles are often harmless, but in some systems they can form sediments which are harmful, such as in spool valves in hydraulic systems. Removal of such very fine particles may require the use of a succession of filters having decreasing nominal pore sizes, and of several different types.

(f) Contamination by water

Water is a very common contaminant in oil systems. It may come from the burning of fuel in a cold engine, or from condensation if the temperature of the system rises and falls in a humid atmosphere. Water can be easily driven off by evaporation if the system operates regularly at temperatures much higher than 100 °C. For this reason engines which normally operate for only a few minutes at a time, on short road journeys or in intermittent generator operation, should be regularly run for longer periods to allow them to warm up thoroughly. Similarly, stationary oil circulation systems should be fitted with tank heaters to ensure that the oil is kept hot enough to drive off water.

If the water remains in the system for only a short time, it may remain clean and cause little trouble. It it remains for a long time in some inaccessible part of the system, then it tends to extract acids and additives out of the oil and become a sort of chemical soup. This mixture will generally be much more corrosive, but will also emulsify more readily with the oil, forming sludges which can block fine holes or filters.

(g) *Microbiological contamination*

In some circumstances the 'chemical soup' mentioned above can lead to another serious problem – that of microbiological contamination. This means simply that microscopic plant life, bacteria, and fungi, start to grow in the oil. In extreme cases this growth can block oil pipes, valves, and pumps, and lead to even greater corrosion. Figure 6.2 shows the appearance of fungus growth in a metalcutting coolant system.

There are three essential conditions for microbiological growth to occur. There must be a source of carbon, and this is supplied by the oil itself. There must be some bacteria or fungal spores present to start the

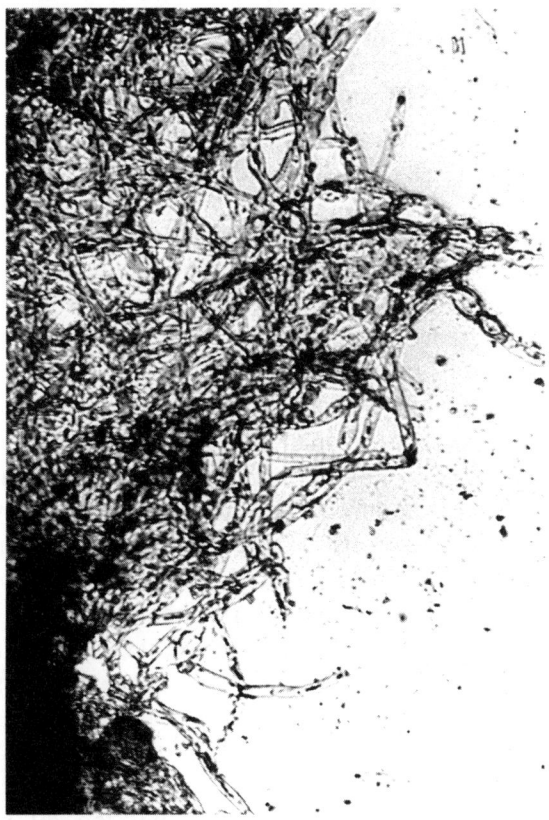

Figure 6.2 Fungus growth in metalcutting coolant

growth, but these are almost universally present in the atmosphere. Finally, there must be free water present, so that the best way to avoid the problem is to keep the system clean and dry. It follows that emulsion lubricants (oil in water or water in oil) are particularly prone to microbiological attack. Two other factors which encourage microbiological growth are a slight acidity in the water (pH about 5 to 6) and slightly raised temperatures (20° to 40 °C), which can give much more rapid growth.

Biocidal additives in the system can prevent microbiological contamination, or kill it once it has started. If a system has become seriously contaminated, then the use of a biocide may kill the organisms and prevent further growth. However, the organic debris which is already present will continue to threaten blockage of the system, and may be very difficult to remove. Prevention is therefore much better than cure.

Systems at greatest risk are large warm systems with low points which can trap water. It is, therefore, particularly important to make sure that all low points in such systems are fitted with drain cocks which are regularly operated to remove water.

The most obvious symptom of bacterial attack in an oil system is often a strong foetid smell of decay. With fungal attack the smell may not be so obvious.

6.4 Establishing the oil change period

Obviously the oil change period will depend on the rate at which the condition of the oil deteriorates and on the amount of deterioration which is acceptable. The specified acceptable limits are only helpful if the oil condition is being monitored in some way. These limits are considered in Chapter 12.

Some form of monitoring of oil quality should probably be carried out on any system containing more than fifty gallons of oil, and on high-speed, hot, or critical systems containing more than about fifteen gallons. In practice these recommendations are rarely followed.

For smaller systems it is probably not economical to monitor the oil, although it should always be remembered that the financial risk if oil is not changed soon enough is the value of the machinery, and not the value of the oil. Where the manufacturer of the system recommends specific oil-change periods, these should be followed unless experience shows them to be unsatisfactory.

It is important to remember that a little observation and common sense can be useful in deciding whether the selected oil change period is right. Any of the following are indications that the oil has seriously deteriorated and should be changed more frequently.

 – Oil black or much darker than when new.
 – Oil flows much less easily than when new.
 – Signs of brown or black lacquer films on moving parts.
 – Sludge or solid debris in the oil.
 – Corrosion of metal parts.
 – Strong or unpleasant smell.

These symptoms may be detectable in a sight gauge or on a dipstick, or possibly only when the oil is drained.

The following additional guidelines may be useful:

(a) well-sealed oil bath systems (i.e., dip, splash, ring or disc lubrication) below about 70 °C; oil-change period 1 year;

(b) well-sealed circulation systems, bulk oil temperature below 50 °C; oil-change period 2–3 years;

(c) well-sealed circulation systems, bulk oil temperature 60°–70 °C; oil-change period 1 year;

(d) well-sealed circulation systems, bulk oil temperature over 70 °C; oil-change period three months, or try to lower temperature;

(e) engines or compressors operating in dirty or humid environment or engines operating intermittently; change more often than manufacturers recommend;

(f) if no top-up is needed, consider changing the oil more often;

(g) if in doubt:

   (i)   look at the colour of the oil in the sight-gauge;

   (ii) if possible, remove the filler cap (*not if the system is under pressure*) and smell the contents;

   (iii) look at the colour, uniformity or viscosity of oil on the dipstick. (Let it cool for a moment before judging whether it is thicker than when new.)

   (iv) drain a small sample from the drain cock, and examine it as for the dipstick; if nothing more, this will probably remove some of the solid debris;

   (v)   have the oil tested, either by some of the simple on-site tests listed in Chapter 12, or by a laboratory.

The problem of disposing of used oil is discussed later in this chapter. In general, however, greater regulation and control of waste disposal is making it more expensive to dispose of waste oils satisfactorily. As a result, major users are finding it economically desirable to keep their lubricant systems in better condition in order to extend lubricant life and thus reduce the cost of disposal, especially of bulk water-containing lubricants such as cutting oils.

Typical of the measures used to extend lubricant life are: improved sealing and screening to keep out contaminants; better filtration and centrifugal cleaning; topping up with fresh oil or additive concentrates; and oil condition monitoring. Most important of all is to improve the awareness of the operators of the commercial benefits of well-maintained lubricant systems.

6.5 Changing the type of oil

It is a good general rule not to change a successful combination. In other words, the type of oil in a system should not be changed unless there is good reason. However, there are several reasons which may seem good; some of these are listed in Table 6.2, together with the nature of the change.

There is always some risk in changing to a new type of oil in a system, and the benefits which are expected should always be balanced against the risk. In the vast majority of cases no harmful effect will arise, but if any fault does occur it is important to recognize it and its causes. In the following, the main risks are described in relation to the type of change which is being made.

(a) *General*

At the present time the vast majority of lubricating oils are mineral oils, and in most cases the change will be from one mineral oil to another. There will therefore be no major incompatibility between the old and new oils.

However, there are a number of synthetic and vegetable oils in use, and serious problems can arise if the oil change involves a change from one class of base oil to another. For example, if a hydroxy-terminated

Table 6.2 Some reasons for changing oil type

Reason for change	*Nature of the change*
1. Marked change in ambient temperature.	Change to more viscous grade if temperature increases, less viscous if it decreases.
2. Oil deteriorates too rapidly, with acidity, lacquering, sludging.	Change to better quality oil with anti-oxidants.
3. Excessive wear or seizure.	Change to oil with anti-wear or EP additives (subject to expert advice).
4. Fires or risk of fires.	Change to fire-resistant oil.
5. Rationalization	Change to equivalent or better quality oil of wider application.
6. High oil consumption with little deterioration.	Change to cheaper oil, or *subject to expert advice* to more viscous oil.
7. Competitive tender.	Change to cheaper oil.

polyglycol is added to a system containing the remains of a mineral oil, the two will not mix or dissolve properly, and the resulting mixture may block oil feeds or filters. Similarly, if a silicone hydraulic fluid is mixed with a mineral oil containing a polymeric viscosity index improver, the polymer may be precipitated out of solution in the mineral oil, and block the whole system.

In practice, the same class of base oil will probably be retained, but it will be important to make sure.

Apart from harmful interaction between the oils themselves, a change to a new class of base oil may have catastrophic effects on rubber seals or hoses, or other non-metallic materials in the system. For example, natural rubber seals can be used with castor oil, but will break down rapidly with mineral oils or most synthetics. Similarly, nitrile rubber seals can be used with mineral oils, but will be rapidly attacked by phosphate esters, chlorinated diphenyls, silicones, and – in some cases – by esters. A change in base oil class may therefore involve changing the non-metallic components in the system to ensure that the materials are compatible.

In most cases, it is good practice to flush out the system after the old oil has been drained out and before the new oil is introduced. In a small system a flushing oil may be used, which is basically a mineral oil without additives and whose viscosity is similar to, or perhaps slightly lower than, the oil used in the system. Even where the difference between the old and new oils is slight, flushing will help to remove the last of the old degraded oil and its contaminants. If the new oil will not mix properly with the old, then it may be necessary to find a flushing oil which mixes thoroughly with both of them.

(b) *Changing to a more viscous grade in summer*

As a general rule, changing to a more viscous grade of the same oil type is the safest sort of oil type change. The only serious risk is that there may be some part of the system where the temperature is near the upper limit which can be tolerated. The effect of changing to a more viscous oil will be to increase the power used in moving the parts of the system, and thus to increase the temperature of the oil. If this effect is added to that of a hotter summer environment, then the temperature may rise to an unacceptable level. The best solution would then be to improve the cooling in the system instead of changing to the more viscous oil.

(c) *Changing to a less viscous oil in winter*

The object of changing to a less viscous oil for a colder climate is usually to ensure ready starting of the system when cold. Once the system is fully warmed up it may not be significantly cooler in winter than in summer. If this is the case, then the oil in the bearings after an oil change may be less viscous than is desirable, and bearing wear or even overheating may take place. When shutting down a hot system, the less viscous oil will drain more completely from the lubricated surfaces, leaving a thinner oil film to cope with the immediate needs of the bearings when re-starting. Finally, the amount of oil lost through leaks and dynamic seals may be greater with a less viscous oil.

(d) *Changing to a better quality oil*

This again is, in general, a very safe type of change. The better quality base oil does not have any harmful effects, and additional additives are rarely harmful. Only occasionally can some unwanted effect arise where a new additive reacts with a non-metal or non-ferrous metal in the system, or with some contaminant which has previously caused no problems. This sort of problem has been reported from time to time, particularly in the chemical industry, where the unknown contaminants are more likely to be reactive chemicals.

(e) *Change to oil with anti-wear or EP additives*

This is an extension of the problem described in the last paragraph. Since anti-wear and EP additives are designed to react with metals under certain circumstances, they may occasionally produce unwanted reactions, with attack on non-ferrous metals or even cast irons or steels. Again, in most cases, no problems will arise.

(f) *Change to fire-resistant oil*

This is the type of change which is most likely to bring serious problems.

The three most common classes of fire-resistant oil are phosphate esters, chlorinated diphenyls, and water-based. The most widely used water-based oils are probably water/polyglycol mixtures and water-in-mineral oil (invert) emulsions. Changing from a mineral oil to any of these oils requires careful consideration of the materials and components of the system.

If it is assumed that in changing the oil the viscosity will not change significantly, there are still a number of chemical and physical properties which must be considered.

(i)   *Phosphate esters*
       The main problem with phosphate esters is that they have a severe solvent effect on non-metallic materials, so that they are perhaps the most difficult class of oils for selecting suitable seals and hoses.

Suitable rubbers are resin-cured butyl, halobutyl, and ethylene-propylene rubber (EPR). A further problem is the rather poor thermal stability of phosphate esters, which restricts them to temperatures preferably below 100 °C.

(ii) *Chlorinated diphenyls*

The stability of chlorinated diphenyls at high temperature is better than that of phosphate esters. Their operating temperature range is, however, limited by their poor viscosity index. This means that they may be unsatisfactory in systems where the temperature varies widely. They attack many plastics, but a suitable rubber for seals is Viton. They also cause some corrosion of copper and copper alloys.

(iii) *Water/polyglycol*

Any water-containing oil must be adequately inhibited to prevent corrosion, and is limited to temperatures between 0 °C and 90 °C.

(iv) *Invert emulsions*

In addition to the limitations mentioned above, emulsions are open to the danger of 'cracking' or separation of water from oil due to the effect of certain chemicals, such as acids, or if frozen. Both invert emulsions and water/polyglycol can be used with nitrile seals.

(g) *Rationalization*

The change to an equivalent or better quality oil which may be recommended for the purposes of rationalizing lubricant supplies carries with it the occasional risks which are listed under paragraphs (b), (d), and (e) above. Where the rationalization study has been carried out by experts, the risk is small.

(h) *Change to cheaper oil*

An oil may be cheaper only because of supply, low transport costs, or any other logistic factor, in which case no special risk is involved. On the other hand, an oil may be cheap because it uses a less highly refined base-stock or because it has a smaller quantity or less effective types of additives. If this is the case there will be increased risk of oil degradation to give acidity, sludging, increased wear, or corrosion. A change to the cheaper oil should therefore always be followed by carefully watching the performance of the oil and the system.

Table 6.3 summarizes the factors to consider with different types of oil change.

Table 6.3 Important factors in changing type of oil

Nature of change	Factors to consider
1. Change to higher viscosity, same class of base oil.	Ensure that any increase in temperature is acceptable.
2. Change to lower viscosity, same class of base oil.	(a) Ensure that lubricant film thickness is still adequate (b) Watch our for leaks (c) Watch out for increased (c) consumption
3. Change to 'better' quality, same class of base oil.	Ensure that additional additives do not introduce new problems
4. Change to cheaper oil, same class of base oil.	Ensure that quality is still adequate.
5. Change to EP oil, same class of base oil.	Watch our for unwanted side-effects of additives, especially on non-ferrous metals.
6. Change to different class of base oil.	(a) Make sure that previous oil is completely removed from system. (b) Ensure that seals and other materials are suitable for new oil.

6.6 Carrying out the oil change

The objective in oil changing is simply to leave the system filled with the replacement oil in as good condition as can be achieved with a reasonable expenditure of money and effort.

Ideally, all the old oil and contaminants should be completely removed from the system. This is rarely practical, for the following reasons:

(a) because of its viscosity, lubricating oil leaves quite a thick film on all the surfaces of the system, bearings, gears, pumps, valves, pipes, etc;

(b) in almost every system there are low points from which oil will not flow to the drain points;

(c) oil is held in filters and in crevices of bolted joints, seals etc;

(d) heavy solid particles, water drops, emulsion, tarry deposits, and so on will stick strongly to the bottom of the reservoir and other surfaces.

To remove these materials completely it would be necessary to strip the system and clean all the component parts. This can only be justified in exceptional cases.

An oil change is therefore normally a compromise between the ideal of complete removal and a reasonable expenditure of money and effort and will involve draining of the old oil with or without some degree of flushing, dismantling and cleaning.

(a) *Draining*

The old oil should be drained as thoroughly as possible by opening all the drain cocks or drain plugs. If a drain cock or plug is difficult to undo, this may indicate that it has seized due to lacquer or tarry deposits (indicating over-heating), hard deposits, or corrosion (due to water or sludge). In such cases it is particularly important for the drain to be opened.

If the system has been static for some time before draining, solid contaminants, water, and sludge may have settled to the bottom of the reservoir so that they will not be removed with the oil. It will be an advantage if the system is drained as soon as possible after operation, so that the oil viscosity will be low and contaminants will be dispersed in the oil. If the whole system cannot be operated immediately before draining, it will help if the oil circulating pump can be kept switched on for some time.

Drainage will also be more efficient if the oil is hot so that its viscosity is low. It will then drain more fully from surfaces, and will be more effective in flushing out contaminants. Operating the system or the oil circulating pump will help to raise the oil temperature, but it may also be possible to use reservoir heaters to bring the oil up to normal operating temperature.

The fraction of the total oil content which can be drained will depend on the design of the system and the viscosity of the oil. It may be as low as 60 percent and will rarely be higher than 90 percent for small systems, or 98 percent for large ones.

(b) *Flushing*

If the system is simply drained and refilled, it will contain something between 2 percent and 40 percent of the old oil, and a similar or higher proportion of the contaminants, probably about 20 percent of both on average. This would be acceptable if the oil were changed before deteriorating too badly, but not if the old oil were seriously degraded or contaminated. If 10 percent of the old solid particles, acid products and water remain in the system, they may be enough to use up most of the additives such as dispersants, anti-oxidants, bases, and corrosion inhibitors contained in the new oil fill.

One solution to this problem is to flush the system out after draining and before refilling for service. If the system is again completely filled with oil, circulated and again drained, a much higher proportion of the old oil and contaminants will be removed, as shown in Table 6.4. In a large system this will not be practical, but flushing with a 25 percent fill may bring considerable improvement, if such a small quantity can be circulated. If between 10 percent and 40 percent of the old oil is still present, then the addition of 25 percent of a flushing oil will bring the total to between 35 percent and 65 percent. Such a quantity may be fairly easy to circulate as well as giving a useful improvement.

The safest and most effective flushing oil is the same oil which is to be used in the system, but this may be too expensive to use for flushing. For this reason most oil companies supply special flushing oils. A typical flushing oil is a plain mineral oil whose viscosity is lower than that of the lubricating oil used in the system. It may be completely without additives or may contain only a small amount of anti-oxidant to control deterioration in storage.

The use of a flushing oil is essential where the replacement oil will not mix freely with the original oil. Expert advice should be taken on the type to be used.

An obvious disadvantage of the use of a flushing oil is that the undrainable oil left in the system will be largely a low viscosity base oil without any of the additives which may be necessary for satisfactory lubrication. In many cases, however, the dilution of the new oil fill by clean base oil will be preferable to the presence of the old oil and its contaminants.

(c) *Dismantling*

Dismantling of the system will usually require a great deal of effort, and will be avoided if possible. Partial dismantling may be carried out in some cases, such as:

(i)   to change a paper or ceramic filter at the same time as the oil is changed;
(ii)   when seals or hoses must be changed to materials compatible with a new class of oil;
(iii)   when failed or life-expired bearings or seals must be replaced;
(iv)   when severe deposits of sludging are known to be present and cannot be adequately drained;
(v)   where heavy bacterial or fungal attack has taken place.

Table 6.4 Effect of flushing on proportion of old oil and contaminants left in system

Percentage of old oil and contaminants left after simple draining (%)	Percentage left after flushing with complete fill and re-draining (%)	Percentage left after flushing with 25 percent fill and re-draining (%)
40	16	25
20	4	9
10	1	2.9
5	0.25	0.8
2	0.04	0.15
1	0.01	0.04

(d) Cleaning

In very large systems, containing 2000 gallons or more, and where heavy deposits, sludging, or contamination often occur, it is impracticable to flush the system. It may then be necessary to clean out the reservoirs and storage tanks by brushing or hosing with industrial cleaning liquids or by steam-cleaning, having closed off the reservoir from the rest of the lubrication system.

Great care must be taken to avoid injury or damage to health from lubricants, contaminants or cleaning fluids when carrying out tank cleaning.

After tanks have been cleaned, it is important to get rid of cleaning fluids and water before refilling the system.

Other parts of the system may be cleaned where necessary, such as when heaters are prone to overheat and become coated with carbonized oil.

(e) Refilling the system

Having ensured that all drain cocks are closed, the main precaution in refilling is to make sure that the fresh oil does not carry contaminants with it into the system. The surfaces of oil drums, measuring cans, funnels, and filling points should be carefully cleaned before refilling begins. In dusty or dirty environments it is preferable for the oil to be pumped from the drum or transport container directly into the filler pipe, using gauze or cloth to filter dust out of the replacement air entering the drum.

In critical systems the replacement oil may be pumped through a high-performance filter before entering the lubrication system. Oil replenishment trolleys are marketed which include a reservoir of replacement oil or hydraulic fluid, a pump, and an efficient filtration system on the delivery line.

Another important precaution is to ensure that the system has been properly filled. It is often possible for an air-lock to prevent oil from passing into some sections of the system during refilling, so that the sight glass or dipstick will show the system to be filled while some sections are still empty. Operating the oil circulation pump, or briefly turning over some of the lubricated components, may help to clear an air-lock. In any case, a careful watch should be kept on sight glasses during the first few minutes of operation to make sure that the required level is maintained.

6.7 Handling the old oil

The procedure for disposing of used lubricating oil depends on the volume and types of oil involved.

In the past small quantities of waste oil, less than 100 gallons, represented a disposal problem. Specialist waste disposal companies and oil re-refiners were generally unwilling to collect quantities of less than 100 gallons, but would accept oil delivered to their premises in smaller quantities. Responsible companies would therefore generally accumulate used oil until a sufficient quantity was available to make it economical to be delivered or collected. In other cases the used oil was often burned or dumped on waste ground, or even poured down drains.

The disposal problem has changed significantly in recent years. There has been a considerable increase in public awareness of the importance of reducing waste pollution. Legislation has also been introduced in many countries to regulate the disposal of dangerous wastes, including petroleum products. Increases in the cost of petroleum products have also made it more economical to collect and process waste oil. This improved economic situation has still not made it worthwhile to collect quantities less than a few hundred litres, but it does mean that the value of the waste oil is now high enough in the United Kingdom and many other countries for the collector and reprocesser to pay for it, instead of having to be paid to remove it, as is still the case in the United States.

Some of the major oil companies have become so concerned about the problem of disposal of small quantities of waste oil that they have set up experimental 'oil banks' in their service stations. Similarly, many local authorities have set up waste oil collection tanks in their civic refuse amenity sites. One report in early 1994 quoted 1354 of them at that time in the United Kingdom, varying in capacity from 600 to 2700 litres, and collecting altogether a total of about 8000 tons of used oil per year.

There is therefore no longer any excuse for waste oil being disposed of illegally by dumping on waste land or poured down drains. Nevertheless, it has been estimated that in the United Kingdom many thousands of tons are disposed of illegally each year by private car owners carrying out their own oil changes. Most of this is believed to be poured down drains, where it causes pollution in rivers and lakes, or damage in sewage treatment works.

Similar problems occur in other countries, although the extent of 'Do-It-Yourself' oil changing may be a particularly British problem. Different countries have introduced different measures to try to control the disposal of waste oil and to increase the proportion available for re-treatment. For example, in Austria, there are restrictions on over-the-counter lubricant sales.

Present estimates are that about 40–50 percent of lubricating oil is irretrievably lost in use, by evaporation, combustion, or leakage (about 20 million tons per year worldwide), and only 25–35 percent recycled (about 10–15 million tons) leaving about 25 percent (10 million tons) lost. There is therefore considerable scope for improvement.

The problem of collection of waste oil by and from major users is of course relatively straightforward. It is now probably normal for any major user to provide waste oil tanks for the collection of waste oils, and in many cases it is possible to sell the oil to a commercial collector. Many factors have combined to make this the preferred procedure in most industries, including fiscal incentives (e.g., in Germany, Finland, and France), the increasing re-sale value of the used oil, the increasing cost of landfill dumping, the pressure of higher environmental standards, and the stringent requirements in many countries for rehabilitation of factory sites.

The one area which is still not straightforward is that of water-containing metalworking lubricants. This is discussed later.

Used lubricating oil is not an attractive material to look at. As a result the tanks in which used oil is collected tend to be regarded as dirty waste bins. They accumulate all sorts of other waste such as cleaning fluids, white spirit, waste fuel, drained cooling water, dirty rags, and even floor sweepings. This converts the used oil from a material with considerable re-use value into a mess which is impossible to re-process and very difficult to destroy safely or even to handle safely.

At the very least, used oil collection tanks should be enclosed and should be filled by means of a large funnel fitted with a wire mesh screen to prevent solid rubbish from being mixed with the oil. A suitable pattern is shown in Fig. 6.3.

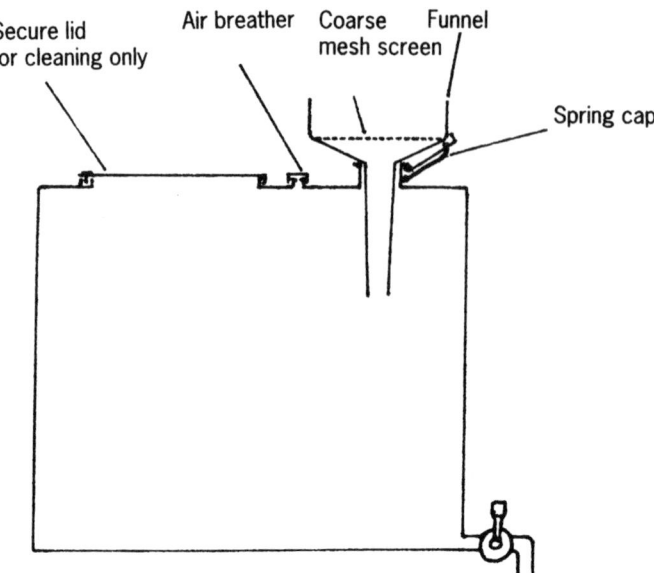

Figure 6.3   Suitable pattern for a waste oil collection tank

Ideally several collecting tanks should be used to collect the following groups of liquids separately:

— used engine oils;
— used mineral oils other than engine oils;
— individual used synthetic oils;
— used emulsion oils and water-contaminated oils;
— fuels and white spirit;
— cleaners and other solvents.

Most of these liquids are flammable, and appropriate precautions must be taken for storing them, especially the fuels, white spirit, and solvents.

Once a suitable collecting system has been established, the disposal problem for small quantities of used oils becomes the same as the problem for large quantities.

Large quantities of used oils, greater than perhaps 500 to 1000 litres, can be treated in one of three ways.

Relatively clean oils, including emulsions and 'soluble oils', in sufficiently large quantities may be 'laundered' or cleaned-up on site to a standard which will allow them to be re-used.

Clean oils and dirty oils can be subjected to a more severe processing to produce fractions suitable for use as fuels or they can be 're-refined' back to lubricant base oil quality for use in manufacturing new lubricating oils.

6.8 Disposing of emulsions and water-contaminated oils

Emulsions are, by definition, dispersions of small droplets of one liquid in another liquid, but they can be 'loose' or relatively unstable emulsions, or 'tight' stable emulsions. Water-contaminated oils may become emulsified, or the water phase may remain mainly or wholly separate from the oil.

One other common class of water-containing lubricant is that of the so-called 'synthetic' coolants used in certain machining operations. These are, in fact, solutions or dispersions of certain chemicals in water.

There are therefore four basic types of water-containing lubricant which may have to be disposed of: loose emulsions, tight emulsions, non-emulsified water and oil mixtures, and solutions of chemicals in water. There are many processes for dealing with these wastes, and the following are some examples.

(a) *Loose emulsions*

The minute droplets of dispersed liquid in an emulsion are prevented from joining together and separating out by electric charges which are produced by surface-active additives or detergents. In a loose emulsion the amount of surface-active material is just enough to produce a stable emulsion, so that the dispersed oil droplets tend to be large, and can be fairly easily made to join together, or coalesce. This coalescing is known as 'cracking' the emulsion. Loose emulsions are sometimes used as rolling oils.

A suitable disposal process for these emulsions consists of the stages outlined below.

(i)   About two percent of a concentrated salt solution is added to the emulsion. This breaks down the electric charges stabilizing the emulsion and allows the oil droplets to coalesce into large drops which can separate out from the water. The salts which are often used include magnesium, calcium or iron chloride, and magnesium, aluminium or iron sulphates. Many others can also be used, and if suitable quantities of a waste salt or acid solution are available from the works it may be possible to make use of them.

(ii) The mixture of oil and water, usually with some solid contamination, is then passed to a centrifuge. By generating a high-speed spinning motion in the liquid, the centrifuge produces centrifugal forces high enough to separate the water and solids from the oil. Alternatively, the treated liquid can be held in settling tanks, where the oil phase separates out on the surface of the water. The oil will usually then be clean enough for disposal with other waste mineral oil.

(iii)   The water may be passed through a separator to remove the last traces of oil and solids. The remaining water must then be treated as necessary to bring it to the standard required for the factory effluent. Where eventual effluent disposal is directly to tidal water, it may be possible to use sodium chloride as the emulsion-breaking salt, as this will usually be acceptable in effluent into sea water.

An alternative process uses an organic demulsifier, such as a polyamine system, which produces less sludge than the salt/acid splitting processes. It may also be cheaper where there is no cheap source of waste salt or acid available.

(b)   *Tight emulsions*

Where there is a higher concentration of surface-active agent, or a more powerful one, the emulsion droplets are finer and the emulsion is harder to break. Soluble oil emulsions used in metalcutting, and invert emulsions used as fire-resistant hydraulic fluids, are generally tight emulsions.

A higher concentration of salt is needed to crack a tight emulsion. This gives a more heavily contaminated water phase, so it will be an advantage to use a process which makes it easier to separate out the chemicals from the water. One suitable process developed by Alfa-Laval is outlined below.

(i)   Two to three percent of a solution of sodium dihydrogen phosphate is added to crack the emulsion. If this is insufficient to give complete cracking, then expert advice may have to be taken.
(ii)   The mixture is centrifuged to separate out oil, water, and solids. The oil should then be suitable for disposal with other waste mineral oil.
(iii) Aluminium sulphate or ferric (iron) sulphate and lime are added to the water to separate out solids as a flocculent precipitate. The precipitate is concentrated in settling tanks, and the bulk of the water passes to the effluent disposal plant. The precipitated sludge is freed of water by means of a centrifuge or filter press, and is suitable for disposal.

This flocculation or coagulation treatment is also suitable for very dilute emulsions or 'synthetic' coolants, where the concentration of oil or other contaminants in the water is very small.

(c)   *Non-emulsified mixtures of water and oil*

These may separate completely on standing in a settling tank, when the water can be drained off from the bottom, and the oil disposed of with other waste oil. The separation is much quicker if the mixture is passed to a centrifuge and this also gives a much cleaner separation.

If the amount of free oil present is very small compared with the total volume of water, it may be possible to separate off the oil by passing the mixture through a hide separator, which will retain the oil.

6.9 Laundering

In many cases the lubricant which is removed from a system will be contaminated but not seriously degraded. In other words the base oil and additives will be almost unchanged after use, but the amount of water, dirt or wear debris will be so great that the oil is no longer fit for use.

In such cases it may be possible to remove the contaminants so that the lubricant is again fit for use in the same type of system. This process is known as 'laundering'. There are two basic techniques used, centrifuging, and filtration or screening. The choice depends on the type of lubricant and on the size and amount of residual contaminant which is acceptable.

For conventional lubricating oils and invert emulsion hydraulic fluids, a high standard of purity is usually necessary, so that solid particles must be removed down to very small sizes. Filters and high-speed centrifuges are both capable of removing particles down to 5 μm or even 1 μm, but because these lubricants have relatively high viscosities, high performance filters are very slow. Centrifuges are therefore preferred, these will also remove contaminating water and sludge. A typical centrifuge is shown in Fig. 6.4.

Soluble oil (emulsion) and synthetic coolants used in machining can become heavily contaminated with swarf as well as with fine metal particles and dirt. Metal screens or rotary strainers have a very high capacity for removing swarf and are continuous and relatively cheap. They may be completely adequate where there is no problem with residual fine particles and where there is no significant contamination with hydraulic fluids or other lubricating oils (known as 'tramp' oil). If necessary screening or straining can be followed by the use of a centrifuge to remove the remaining fine particles and any free 'tramp' oil.

Most cleaning processes involving screening, centrifuging, or even settlement on standing, will be much more effective if the liquid is heated. This reduces the viscosity, improving the overall flow and allowing the various phases to separate more easily.

Effective removal of solid contaminants and tramp oil on a regular basis is the best possible way to treat used metalcutting coolants/ lubricants. This can extend the life of the lubricant by a factor of five or more, thus reducing the waste disposal problem by the same factor.

Technically, laundering means simply the removal of contaminants

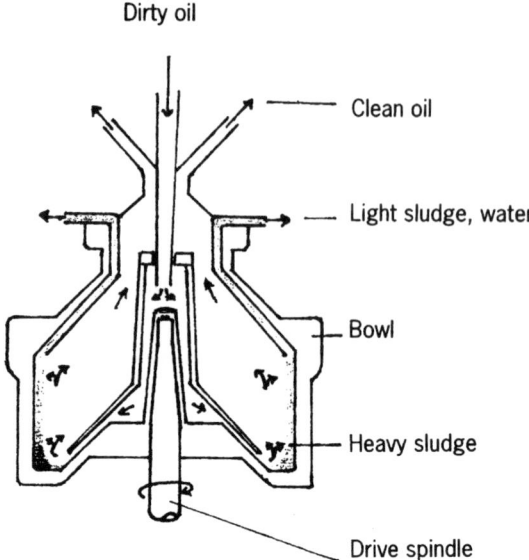

Dirty oil

Clean oil

Light sludge, water

Bowl

Heavy sludge

Drive spindle

Figure 6.4 Typical centrifugal oil cleaner

from the lubricant, but it may also be useful to make up some depleted component in the oil. The simplest example is the addition of fresh oil concentrate to a soluble oil coolant. The oil droplets in the soluble oil emulsion will stick to metal surfaces or be extracted out into tramp oil, so that their concentration slowly decreases. If this happens the required quantity of fresh oil concentrate can be added to bring the concentration in the coolant back to its proper level.

6.10 Re-processing for fuel use

Laundering techniques can only remove contaminants which are not dissolved in the oil, such as metal and other solid particles, water (in straight oils or invert emulsions) and oil (in soluble oil emulsions or synthetic coolants). Other contaminants may be present which dissolve completely in the oil, such as fuels or light solvents, other oils, and oxidation products from the oil and its additives.

Waste oils from which water and solid contaminants have been removed *and which do not contain any light fuels or solvents* are often blended into the works fuel oil supply and burned. They produce almost the same amount of heat as the fuel oil and thus produce a straight volume-for-volume saving in the fuel bill.

However, there are several important limitations on the use of waste oil as a fuel or fuel supplement. The waste oil may contain a flammable liquid which will create a fire or explosion hazard, or may contain a non-flammable substance such as a chlorine-containing solvent which will interfere with burner operation. Less obviously it may contain soot-forming contaminants such as toluene or coal-tar naphtha, which will soot up burners. Finally, most waste oils, and especially waste engine oils, contain additives which produce undesirable pollutants in the smoke after burning.

Burning of untreated waste oil is therefore usually restricted in the United Kingdom to workshop space heaters of less than 400 KW. The pollutants of greatest concern in flue gas are polychlorinated biphenyls, polynuclear aromatic hydrocarbons, heavy metals, and dioxins (formed on burning certain chlorinated compounds). The concentrations of some of these can be reduced by mild treatment of the waste, such as with alkali to remove acidic products, or with absorbents to remove some of the chemically active or marginally soluble compounds.

Flue gas contaminants from large installations, such as 1 MW or more, are usually monitored. The permissible concentrations vary for different contaminants and in different countries, but may be as low as 20 ppm.

Where the composition of the waste oil cannot be improved enough for these limits to be met, even with blending, the number of possible outlets is severely reduced.

Destruction of polychlorinated biphenyls and other chlorinated compounds without the production of dioxins requires incineration at 1400 °C or more. This is the last resort as a method of disposal, because not only is there no economic return, but there will also be a charge for disposal in this way.

Cement kilns operate at temperatures up to 2000 °C, so that provided the metal content of the waste oil is not too high, it can be used as fuel in cement kilns without any pollution hazard. However, the potential market for this is limited.

One market which is not yet fully exploited is use in association with scrap vehicle tyres in a gasification process to produce 'synthesis gas' or 'syngas'. This is a mixture of light hydrocarbons used in chemical synthesis plants. Because the product is not released to atmosphere, emission controls as such do not apply. Composition restrictions depend on the use to which the syngas is put, but various cleaning processes such as precipitation and scrubbing can be used. At present the number of gasification plants is very small, but the number may increase in future.

There is therefore a tendency to divert much of the remaining oil for use as fuel in situations where flue gas composition is not restricted. As the restrictions become tighter and more widespread, this means that more and more waste oil is being used as marine furnace fuel.

The one remaining disposal process is re-refining, in which the waste oil is converted to lubricant grade base oil.

6.11 Re-refining

In theory the treatment of waste oil to regenerate lubricant base oil is the ideal conservation process, and in fact European Union directives call for priority to be given to re-refining in the disposal of waste lubricating oil. In spite of this, the amount of waste oil currently re-refined in Europe is only about two percent of the amount of new oil manufactured, which is probably similar to the proportion elsewhere in the world.

The conservation argument is not simple. Waste oil which is used as fuel replaces an equivalent volume of new fuel oil, while re-refining requires a significant energy input.

There are many different re-refining processes in current use. The traditional process is the acid/clay process, and on a world-wide basis this is still the most widely used, because over 2000 acid/clay plants are believed to be still in use in China.

The schematic arrangement of a typical acid/clay process is shown in Fig. 6.5. The waste oil is transferred to a large storage tank through a wire screen, which removes large solid objects. The oil is usually allowed to stand in this tank for several days, after which the bulk of the water and sludge can be drained off from the bottom. Either heating or the addition of alkali may be used to improve the separation.

The next stage is usually to remove light fuels and solvents in a vessel in which live steam is passed through the oil. The steam carries away with it any contaminating petrol, paraffin, or diesel fuel. The mixed vapour is condensed and the fuel fraction can be separated fairly easily from the water.

The oil is then heated to 40°–50 °C and contacted thoroughly with concentrated sulphuric acid. The acid reacts with most of the non-hydrocarbon substances and extracts them from the oil, producing an acid tar which is run off from the bottom of the tank. The oil is then

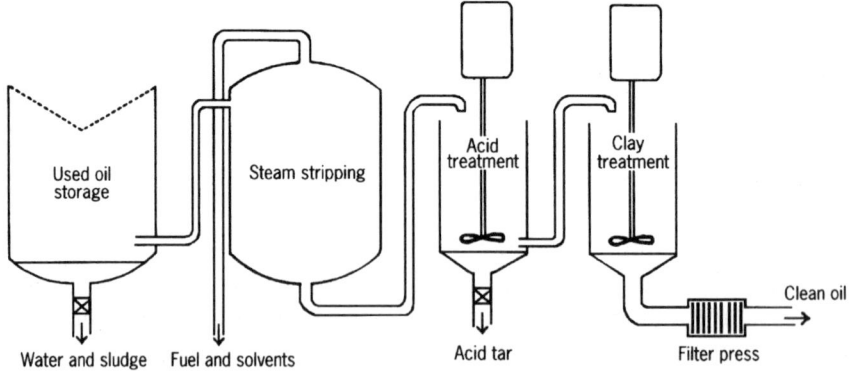

Figure 6.5 Typical acid/clay re-refining process

Table 6.5   Commercial re-refining processes

Name	Primary treatment	Finish
Meinken	Heat, sulphuric acid	Clay absorption
IFP	Liquid propane extraction	Clay absorption
Viscolube	Solvent extraction	Vacuum distillation/clay
Viscolube	Thermal deasphalt	Vacuum distillation/clay
RTI	Cyclonic vacuum distillation	Clay absorption
LUWA, Pflauder	VTFD*	Clay absorption
KTI Relube	VTFD	Hydrotreatment
Evergreen	Chemical treatment, VTFD	Hydrogenation
DEA/Edelhoff	VTFD/solvent extraction	Hydrofinish
Snamprogetti	Solvent extraction, vacuum distillation	Hydrotreatment
Leybold/Recyclon	Sodium treatment	Vacuum distillation
Phillips Petroleum	Demetalize, filter	Hydrotreatment

*VTFD = Vacuum thin film distillation

filtered through a bed of Fuller's Earth to remove any remaining acid and solid particles.

The major difficulty with this acid/clay process is that of disposing of the acid tar and acid-contaminated clay (Fuller's Earth) which are produced. Traditionally these have been dumped in landfill sites, but such disposal is no longer acceptable in most countries. As a result a number of alternative processes are now in use which generate much lower quantities of waste. Some of the most important are listed in Table 6.5.

Acid/clay plants can be economical in relatively small capacities, 1500–5000 tons per year, whereas most of the higher technology plants require a throughput of 30 000 to 50 000 tons per year. It seems likely that the extent of re-refining will increase in future, and the volumes available will encourage the use of the larger plants.

There has been argument for many years about whether the quality of re-refined base oils can match that of fresh refined oils. The difficulty of removing certain contaminants, especially polycyclic aromatic hydrocarbons, is claimed to be offset by the fact that the most unstable components of the base oil itself have been removed in the course of its previous use.

One valuable proof of the quality of re-refined oil is that automotive oils formulated from re-refined base stock have been given CG–4/SH API quality approvals in the United States, and Mercedes-Benz, Volkswagen and MIL–L–46152B approvals in Germany. It is also possible that the very stringent Dexron III approval has now also been achieved.

Chapter 7

Greases and Anti-seizes

7.1 The nature of greases

Lubricating greases are not simply very viscous lubricating oils. They are lubricating oils in which thickeners have been dispersed to produce a stable 'colloidal' structure or gel, similar in some respects to an emulsion.

In previous chapters it was explained that an emulsion is a dispersion of small droplets of one liquid in another. The small droplets are prevented from joining together and separating out by the presence of static electrical charges on the droplets. The electrical charges are maintained by surface-active substances which concentrate at the interface between the droplets and the continuous liquid phase.

A concentrated emulsion can have a viscosity far higher than either of the two liquids which form it. The increase in viscosity is caused by the fact that the stabilized droplets resist any deformation from their spherical shape. However, if the droplets are close together the surrounding continuous liquid cannot flow easily between them.

The same is true of dispersions of solid particles in a liquid if the solid particles are prevented from sticking together and settling out by electrical charges in the same way as the droplets in the emulsion. Such a stabilized dispersion of fine solid particles in a liquid is called a colloidal dispersion. When the dispersion as a whole is concentrated enough to behave as a solid it is called a gel.

It is important to understand this aspect of grease structure because it affects much of the behaviour and many of the important properties of the grease.

7.2   Composition of greases

The liquid phase in a lubricating grease is always a lubricating oil. Mineral oils are again the most widely used, and probably over 98 percent of greases are made with mineral oils. Synthetic esters and synthetic hydrocarbons come next, and most of the remainder use silicones. Recently, there has been a growing interest in fire-resistant greases for high-risk applications, and phosphate ester greases have been developed. For very high temperatures (over 200 °C) greases made from perfluoropolyethers are also available, but these are extremely expensive and are only used in very small quantities. Greases based on vegetable oils have been developed for use in the food industry but these too are not widely used.

The viscosity of the liquid used in making a grease is important. It has some influence on the consistency (i.e., softness or hardness) of the grease, but this is more dependent on the amount and type of thickener. The thickener can be altered to produce a grease of the required hardness even if the viscosity of the base oil varies.

The effect of oil viscosity on the performance of the grease is much more complicated than just to affect the consistency. Some major users will specify both the consistency of the whole grease and the viscosity of the base oil. This subject is discussed further in Section 7.5 (a).

Some of the base oils used in grease manufacture are listed in Table 7.1.

The second most important component of a grease is the thickener, and grease types are usually classified by the type of thickener used. The

Table 7.1   Some components used in grease formulation

Base oils	Thickeners	Additives
Mineral oils	Sodium soap	Anti-oxidants
Synthetic hydrocarbons	Calcium soap	Anti-wear additives
Di-esters	Lithium soap	EP additives
Silicones	Aluminium soap	Corrosion inhibitors
Phosphate esters	Lithium complex	Molybdenum disulphide
Perfluoropolyethers	Calcium complex	Friction modifiers
Fluorinated silicones	Aluminium complex	Metal deactivators
Chlorinated silicones	Bentonite clay	VI improvers
Polyglycols	Silica	Pour point depressants
	Carbon/graphite	Tackiness additives
	Polyurea	Water repellants
	PTFE	Dyes
	Polyethylene	Structure modifiers
	Indanthrene dye	
	Phthalocyanine dye	

most common thickeners are soaps of calcium, sodium, lithium, or aluminium, (and less often lead, barium, or strontium), either alone or associated with other components. Others include bentonite clay, silica, carbon black, and several different polymers.

Because soap-based greases are so important it may be worth explaining briefly what these soaps consist of.

Animal fats and vegetable oils are esters, theoretically formed by the reaction between glycerol (glycerine) and long-chain fatty acids, as shown in Fig. 7.1. If they are heated with an alkali such as lime or sodium hydroxide (caustic soda) they break down to give free glycerol and a mixture of the alkali metal salts of the various fatty acids present, as shown in Fig. 7.2. This process is called saponification, which simply means conversion into soap.

The product is a mixture of soap, glycerol, and water. If the water is removed the residue is a 'soft soap'. Such soaps are very effective in thickening lubricating oils to produce greases. For special applications a single specially prepared long-chain fatty acid may be used to prepare the soap, the most widely-used being 12-hydroxy-stearic acid.

Unlike the dispersed droplets in emulsions, the thickener particles in greases are not spherical. Those in soap-base greases are, in fact, fibre-shaped – as shown in Fig. 7.3. These can give a more or less fibrous texture to the grease as a whole. The interlocking of the fibrous thickener particles probably helps to give a harder consistency to the

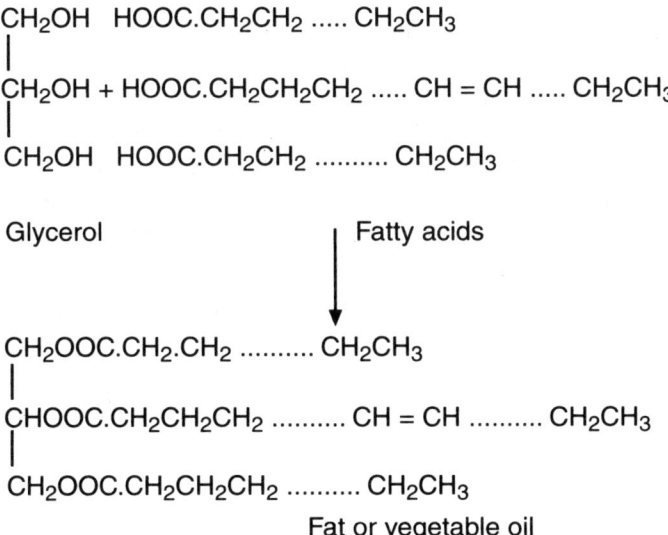

Figure 7.1 **Theoretical mechanism for formation of animal fat or vegetable oil**

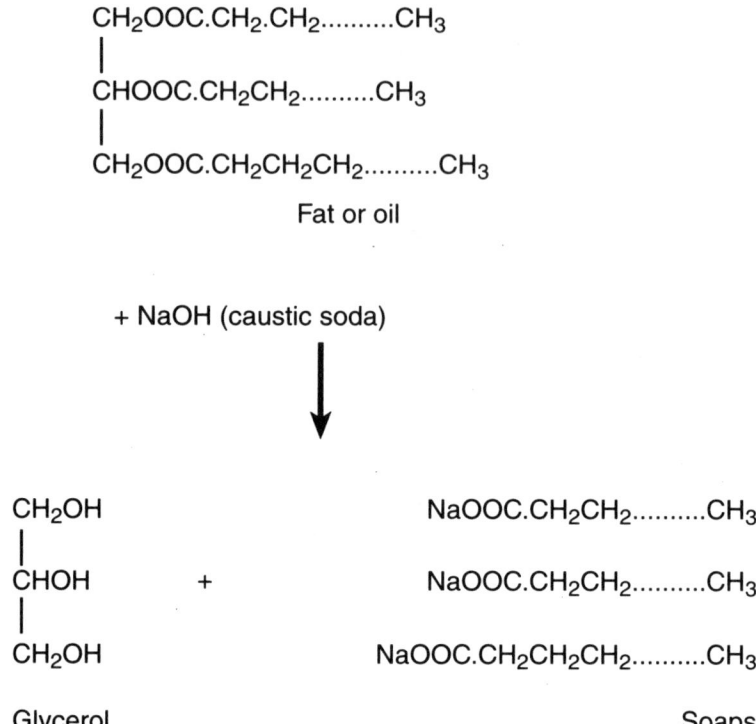

Figure 7.2   Saponification of fats and oils

grease. It is found that non-soap thickeners, which are not fibrous in shape, must often be used at higher concentration than soaps in order to give the same degree of thickening. Some of the thickeners used in grease manufacture are listed in Table 7.1.

Some of the additives which are used in greases are similar to those which are used in lubricating oils, including anti-oxidants, corrosion inhibitors, and EP or anti-wear additives. In addition, friction modifiers and structure modifiers may be used. Friction modifiers include PTFE powder, nylon, mica, graphite, and molybdenum disulphide (which also acts as an EP agent). Structure modifiers are substances which affect the interaction between thickener and base oil, and they include latex and several different polymers.

Some of the additives used in greases are listed in Table 7.1.

7.3   Grease manufacture

There are important differences in the manufacturing processes for different greases, depending on the type of base oil and thickener and on the softness or hardness required in the finished grease. The general procedure is to disperse the thickener in a small quantity of the base oil

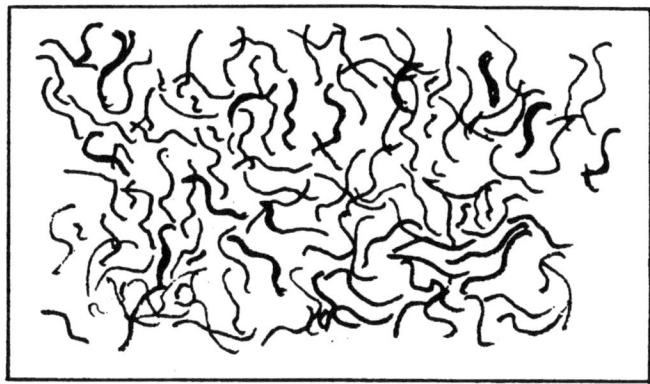

Figure 7.3   Fibrous structure of soap greases

at high temperature, and then to add the remainder of the base oil slowly with stirring as the dispersion cools. Additives are usually added in the last stages of dilution and cooling.

In the manufacture of soap-based greases the fat or fatty acid will usually be dispersed in the small quantity of hot base oil, and the alkali added to saponify it. In this way the saponification reaction helps to disperse the resulting soap in the oil. After all the base oil and additives have been added, the grease may be subjected to a milling operation to give a smoother consistency before packaging.

With non-soap greases it is sometimes necessary to disperse the thickener in the whole of the base oil, and a stable gel structure is obtained by carefully controlled milling. In other cases the thickener is dispersed in a small part of the heated base oil and the hot slurry is milled to improve dispersion before combining with the bulk of the oil.

Although some grease is manufactured in continuous plant, batch manufacture in large 'kettles' is more often used. The manufacturing process is sensitive and hard to predict, especially where the saponification process is carried out from natural fats, or where some of the more unusual non-soap thickeners are used.

The manufacture of greases is more of a craft than a science, and even today is often described as witchcraft. The industry is becoming more scientific, but so far the improvements have been concerned mainly with much more careful control, monitoring, and recording of the various materials and processes. By these means it is now usually possible to produce consistent batches of grease once a satisfactory process has been established, but it is still not possible by scientific means to predict the properties and performance of a new grease formulation without experimentation.

Where careful control and monitoring are not used, there will often be variations in successive batches produced by the same nominal process. It is then usually possible to correct a batch to the required consistency, but the basic structure of the grease will be variable. As a result, the performance is also likely to be variable, especially the operating life and even the storage life.

7.4 Mechanism of action of greases

It is sometimes said that a grease in use acts as a spongy reservoir of base oil, releasing oil slowly to flow onto the bearing surfaces. Some studies with grease-lubricated ball bearings and disc test machines have indicated that this is not the case. After some hours of operation the system reaches equilibrium, with the bulk of the grease in the bearing covers and a very small, apparently liquid, amount actually on the disc, ball, or race surfaces. On analysis the material on the contacting surfaces is found to contain much the same concentration of thickener as the original grease. The concept of lubrication being performed by oil released from the grease is therefore not correct.

In fact, grease as a whole is an effective lubricant and in spite of its firm consistency it will flow satisfactorily in bearings, gears, and other moving systems.

Certainly in grease-lubricated plain bearings, gears, gear couplings, splines, chains, etc., grease flows as a whole, and there is no separation of oil. On the other hand, grease in storage often suffers some 'bleeding' or slight separation of oil from the bulk. It seems possible that where there is a quantity of undisturbed grease present in a system, such as in the covers of ball bearings or in the unswept part of a gearbox, some separation of oil may take place which can assist in lubrication. But this is not the main mechanism of action of the grease, and is probably not essential for lubrication.

The amount of grease which is required for effective lubrication at any one instant in a system is very small; it is just enough to ensure that a film is maintained on the contact surfaces. It is normal practice to supply a far greater quantity of grease, especially where there is no re-supply arrangement.

When the system starts operating the surplus grease is quickly swept clear of the region which is occupied by the moving parts (balls, rollers, gears, etc.). If its consistency (i.e., degree of hardness) is correct it will remain clear of this swept volume. If it is too soft, so that it slumps or collapses back into the path of the moving parts, then it will be continually churned, and this churning can generate a great deal of heat, which may be enough to destroy the grease. The same problem can arise if too much grease is present, so that it cannot all be accommodated in the space available clear of the moving parts.

The consistency and quantity of grease used in a system are therefore quite critical, and both factors will be considered later.

The surplus grease in the system serves two useful purposes. (a) It serves as a reservoir to maintain the small quantity which is actually involved in lubricating the contacting surfaces; (b) it can form a very effective seal against loss of lubricant from the system, and against ingress of dirt or other contaminants into the system.

There is some disagreement about the reservoir effect. It seems certain that in most industrial ball bearings the grease life is much longer if this surplus is present. On the other hand, in some special applications such as spacecraft bearings comparable lives have been obtained by a technique called 'grease plating' in which only a thin film is applied to the contact surfaces of the bearings. There are therefore three points of view about the beneficial effect of the surplus grease in an industrial ball bearing.

(a) The surplus provides a reservoir from which the base oil, containing the additives, slowly feeds to the working surfaces. The objections to this as an important effect were described earlier.

(b) The surplus acts as a reservoir from which additional grease feeds to the working surfaces as necessary. One theory is that if the film of grease on the working surfaces decreases there will be a slight increase in frictional heat. The surface of the surplus grease close to the moving balls then becomes hot, expands, and softens and smears extra grease onto the ball surfaces.

(c) The surplus does not act as a reservoir, but only as a seal, preventing the lubricating film from evaporating or migrating out of the bearing, and thus extending its life.

Whatever the mechanism, there is no doubt that the surplus grease increases the life before relubrication of the bearing becomes necessary.

The fact that in grease lubrication the lubrication is only performed by a thin film of grease on the contact surfaces has several important consequences. Practically all the friction in the bearing is converted to heat, and this heat cannot be dispersed in a relatively large volume, as is generally the case in oil-lubricated bearings. Even where there is a large volume of grease nearby in the bearing covers or a gearbox, no flow takes place in the grease, so that there can be no convective cooling. Heat can only be removed through the shaft and the bearing housings, and this is relatively inefficient.

As a result a grease-lubricated bearing tends to run hotter than an oil-lubricated bearing under the same conditions of speed and load. The higher temperature leads to greater oxidation or thermal breakdown, and the breakdown products are retained in the very small amount of grease in the running film, instead of being dispersed in a greater volume. It follows that grease cannot be used at speeds as high as for oil because the frictional power loss leads to higher temperatures and more rapid deterioration of a grease.

7.5 Properties of greases

(a) *Consistency*

The most important property of grease, which distinguishes it from other types of lubricant, is its semi-solid nature. Greases can in fact vary from very soft, or semi-liquid (rather like thick cream) to hard (like a block of wax). The relative hardness or softness is called the consistency.

Consistency is measured in terms of penetration – the distance in tenths of a millimetre to which a standard metal cone under standard conditions will penetrate into the surface of the grease. The penetrations almost all lie between 475 for a very soft grease and 85 for a very hard one. Consistency is also commonly described by the NLGI classification developed by the American National Lubricating Grease Institute. The relationship between NLGI number and penetration for the full range of greases is shown in Table 7.2.

The consistency of a grease is important in determining its suitability for particular applications. Consistency can, however, change if a grease is worked in a bearing or gearbox, or pumped through a pipe. The extent to which the consistency can change due to mechanical working depends on the stability of the gel structure. An unstable grease may become very soft, or even flow like a liquid when heavily worked, but more often there is a slight softening or sometimes a slight hardening.

Since a grease is very soft compared with most solids, or perhaps because its function is similar to that of an oil, it tends to be thought of as a very viscous liquid rather than a soft solid.

The difference between a very viscous liquid and a very soft solid is

Table 7.2 NLGI grease classification

NLGI number	Worked penetration at 25 °C
000	445–475
00	400–430
0	355–385
1	310–340
2	265–295
3	220–250
4	175–205
5	130–160
6	85–115

that if a pressure is applied to a liquid it will start to flow, however small the pressure. With a very low pressure and a very viscous liquid the flow rate may be minute, but it will not be zero, and the liquid will not revert to its original shape when the pressure is removed. When a solid is subjected to a very low pressure, it will deform elastically, so that if the pressure is removed the solid will recover its original shape. As the pressure is increased there comes a point where the elastic limit is reached. At higher pressure the solid will yield, or deform plastically, which means that it will flow. The critical pressure at which it starts to flow is called yield stress.

Greases act as something between viscous liquids and soft solids. Because of its consistency a grease offers a very high initial resistance to flow. Once moving, and being sheared at high speed between the surfaces of a bearing or gear, its resistance to flow is very much reduced, and may be determined by the viscosity of the base oil more than by the consistency of the grease.

It is possible to have a viscous oil whose resistance to flow is greater than that of a typical grease. An extreme example would be a silicone oil with a viscosity of 500 000 cSt, which would probably be 1000 times more resistant to flow than a grease. If such an oil could be made to flow at high speed in a bearing, its viscosity (apart from temperature changes) would still be 500 000 cSt so that its resistance to flow would still be enormous. It would, in fact, behave more like an adhesive than a lubricant. With a grease however the initial resistance to flow is high, but when it is rapidly sheared in a bearing its apparent viscosity may drop to that of its base oil, perhaps only 100 cSt at 40 °C.

The viscosity of an oil changes quite rapidly with change in temperature. The viscosity of a very viscous oil might decrease by 80 percent if the temperature increases by 25 °C. The consistency of a grease changes relatively little with temperature until it reaches its drop point (see below).

These differences in flow properties between greases and oils are the main factors affecting the choice of grease or oil for the lubrication of a particular system.

Figure 7.4 shows how the apparent viscosity of a grease decreases as the rate of shearing increases.

(b) *Effects of temperature*

If a soap-based grease is heated, its penetration increases only very slowly until a certain critical temperature is reached. At this point the gel structure breaks down, and the whole grease becomes liquid (see Fig. 7.5). This critical temperature is called the dropping point or drop point. Typical drop points for a variety of greases are listed in Table 7.3.

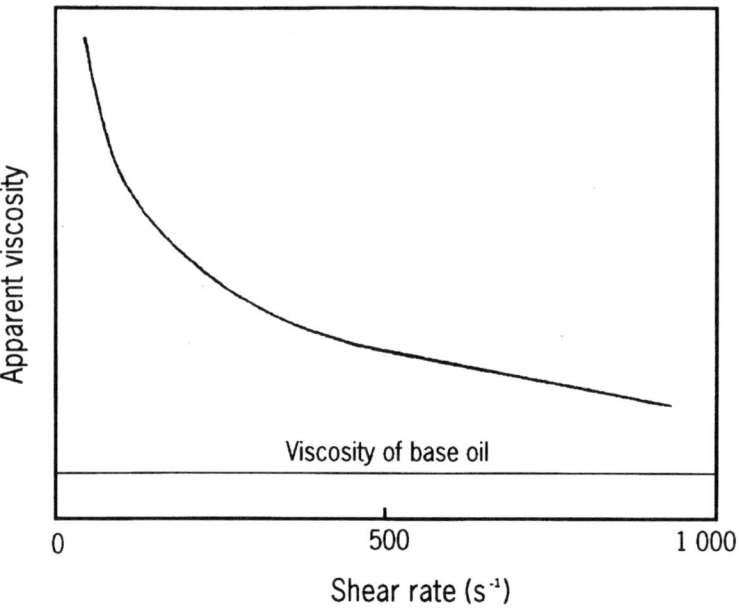

Figure 7.4   Change of apparent viscosity with shear rate

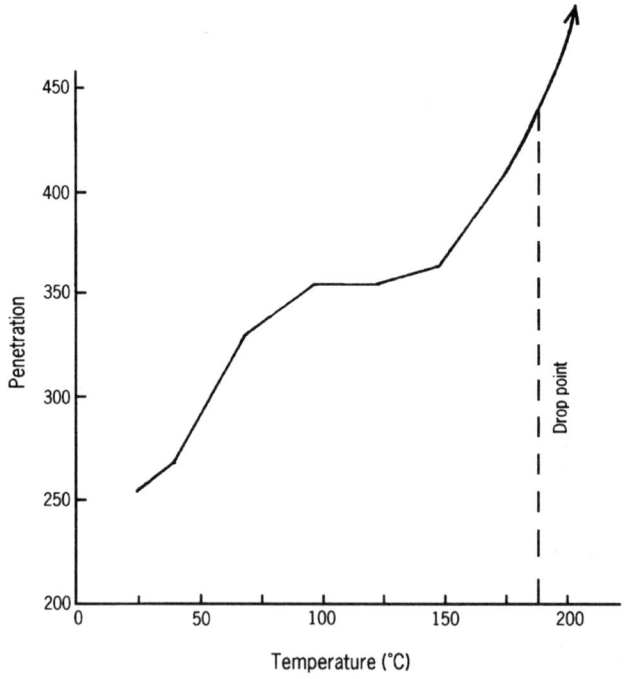

Figure 7.5   Variation of penetration with temperature for a sodium soap grease

Table 7.3 Effect of thickener on drop point

Thickener	Drop point (°C)
Sodium soap	165–190
Barium soap	120–150
Calcium soap	60–100
Aluminium soap	100–120
Lithium soap	170–200
Calcium complex	225–250
Aluminium complex	200–260
Lithium complex	250–320
Bentonite clay	210–250 (breakdown)
PTFE	230–300

Some non-soap greases do not show a drop point, and the structure remains stable until the temperature rises to the point at which either the base oil or the thickener decomposes.

If a grease is heated above its drop point and then allowed to cool, it is commonly found that it does not regain the same consistency. Subsequently, its performance will often be unsatisfactory. The drop point is therefore an important, though not an absolute, limit to the temperature at which a grease can safely be used.

The importance of the drop point as a limit on the operating temperature depends on the form in which the grease is present in a mechanism, and the way in which it is behaving.

Where the grease is present in an undisturbed situation, such as in the covers of a rolling bearing or the free space in a gear-box, it usually has one or two functions. The first of these is to provide a seal to keep dirt out of the mechanism, or lubricant in. The second is to provide a reservoir from which the working grease on the lubricated surfaces may be replenished. In this situation the drop point is a very important, or even an absolute temperature limit, because heating the grease above its drop point may result in a permanent change in consistency. Such a change could affect the sealing performance and the re-supply performance. Even more important is the fact that an increase in penetration (that is, a softening) could result in the grease 'slumping' into the moving parts, leading to churning and over-heating.

Similarly, if the grease is present in bulk in a moving mechanism such as a spline, universal coupling or low-speed gear, its consistency is critical in ensuring that it feeds correctly onto the contact surfaces but does not escape from its desired location. Exceeding the drop point is potentially damaging because of possible softening.

Where the grease is present in the thin lubricating film on moving contact surfaces, especially in high-speed contacts, its structure is already disrupted, and exceeding the drop point is relatively unimportant.

The other important factor which limits the high temperature use of a grease is oxidation.

With decreasing temperature a point is ultimately reached where a grease is too hard for the bearing or other greased component to be used. This represents the lowest operating temperature for the grease, although there is some variation in this limit, depending on the shape of the component and the power available.

The low temperature behaviour depends mainly on the base oil, and greases for low temperature use must be made from oils having a low viscosity at that temperature.

The base oil in a grease will oxidize in exactly the same way as a lubricating oil of similar type. The thickener will also oxidize, but in most greases the thickener is less prone to oxidation than the base oil.

If the grease is static, as in the covers of a bearing, the rate of oxidation will be slower than in a similar oil, because the solid nature of the grease interferes with access of oxygen to the interior. If the grease is being actively worked the oxygen access is improved, and because of the heat generated by working oxidation can be rapid. Oxidation in the thin lubricating film on the moving contacts is therefore an important factor limiting the usable temperature.

When grease oxidizes it usually darkens, and there is a build-up of acidic oxidation products, just as in a lubricating oil. These products may have a detrimental effect on gel structure, and cause softening, oil bleeding, and leakage. Since grease does not flow readily, heat transfer is poor, and serious oxidation can begin at a hot point and spread slowly through the grease. This produces carbonization and progressive hardening or crust formation, and overall the effects of oxidation tend to be more harmful in a grease than in an oil. The breakdown products are particularly damaging in the thin lubricating film on the moving contact surfaces.

One special case is where the base oil is a polyglycol. Here there is no tendency to carbonization and the polyglycol progressively volatilizes away as it becomes overheated, leaving only thickener. In order to ensure that the thickener residue is as effective a lubricant as possible, it is common practice to thicken a polyglycol grease with carbon or graphite.

One curious effect which is not entirely understood is that the maximum temperature for use of a grease is lower by 20 °C or 30 °C in very small ball bearings than it is in larger ball bearings. This may be due to the greater surface-to-volume ratio in small volumes, allowing proportionally greater oxygen access.

The temperature limits for use of greases are therefore determined by drop point and oxidation, and by stiffening at low temperature. These effects are summarized for certain greases in Fig. 7.6.

(c) *Migration of base oil*

The base oil in a grease can flow through the thickener network under certain circumstances, as if it were passing through a porous medium such as a filter. This property is known as permeability. It does not seem to occur in most normal applications, and does not usually affect the behaviour of a grease. However, there are two situations in which it can be important.

The first arises when grease is pumped through a pipe, such as in a centralized grease lubrication system. If for any reason there is an abnormal resistance to flow at some point in the system, then the grease may form a plug with pressure building up behind it. When this happens the base oil in the grease behind the plug may start to flow through the plug, giving free oil or softened grease ahead of the plug, and leaving excess thickener behind which will intensify the plugging. Greases which are to be pumped through long pipes should therefore be tested, such as in a pressure filter, to ensure that they do not behave in this way.

The second situation is when two different greases are present in the same system. Where the two greases are in contact their base oils may migrate across the interface, so that a certain amount of exchange takes place. The base oil and thickener in each grease will have been carefully matched in manufacture to give the required consistency and flow properties. Exchange of base oils can therefore result in poor matching of base oil and thickener with the result that one or both of the greases breaks down and liquefies or separates.

(d) *Compatibility*

This effect of base oil exchange is one aspect of the problem of compatibility. In most cases, when two greases of similar type are mixed, no serious problem arises, but occasionally incompatibility leads to complete failure. As a general rule, therefore, two different greases should not be mixed in a system. Where nipples are fitted for re-greasing a bearing and the type previously used is not known, it is good practice to feed fresh grease until all the old grease is flushed from the bearing and fresh grease appears at the outlet.

The compatibility of greases with rubber seals and other non-metallic materials is also important. This depends mainly on the base oil and has been considered in Chapter 3.

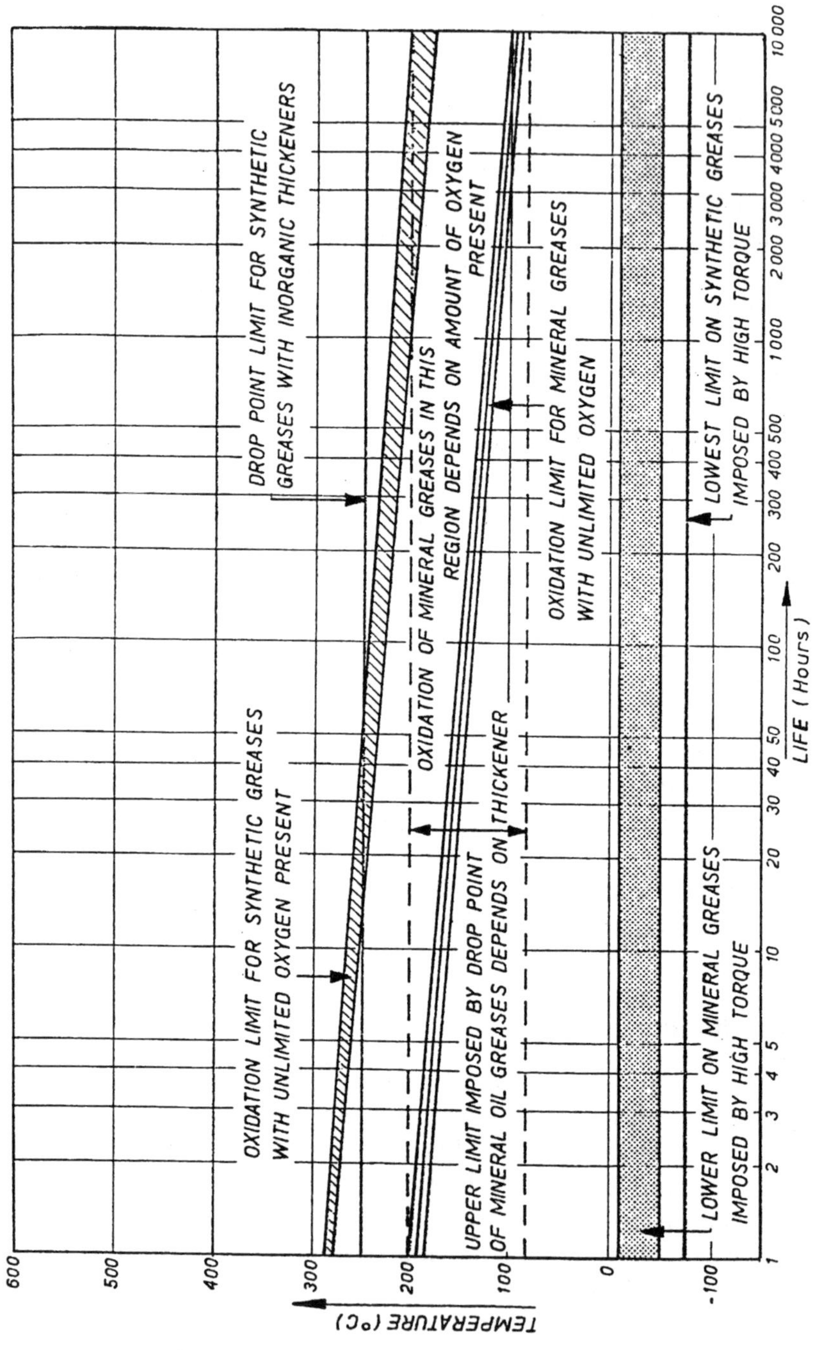

Figure 7.6 Temperature limits for some greases

(e) *Contamination*

Greases can become contaminated with wear debris, dirt, breakdown products, water, and other liquids, in the same way as lubricating oils, but the effects in greases are usually different.

In an oil-lubricated system solid contaminants can be flushed out from the bearing surfaces and deposited harmlessly in some other part of the system. With grease this cannot occur, and solid particles remain held in the grease. Thus wear debris remains in the contact zone, and grease which is fed into the contact zone will carry with it any dirt which contaminates it. This also applies to any debris formed by overheating or other deterioration of the grease. It is therefore particularly important to make sure that the grease fed into a system by means of a grease gun or centralized system is kept clean and free of dust or other contaminants.

On the credit side, the surplus grease in the covers of a ball bearing or inside the case of a gearbox or other device can trap dirt and prevent it from reaching the contact surfaces.

If a grease becomes contaminated with oil, fuel, or a solvent in which the base oil can dissolve, then the grease structure will break down and the grease may become liquid. Liquids in which the base oil is not soluble are less harmful but can slowly damage the gel structure of the grease and soften it. Water can act in this way, initially softening the grease, then dissolving some soap thickener and emulsifying base oil, and finally washing away some of the degraded material. Non-soap thickeners such as bentonite clays or silica will not dissolve in water, but the gel structure may still be made unstable so that the grease softens.

Acidic oxidation products can also de-stabilize the structure, causing softening, but ultimately oxidation tends to cause hardening of the grease. When fresh grease is fed into a bearing which has been hot, the acidic oxidation products in the old grease may attack its structure, thus causing rapid breakdown. As far as possible sufficient fresh grease should be introduced to flush out all the old degraded grease.

7.6 Advantages and disadvantages of grease

The main function of an oil or a grease is to interpose a film between two surfaces so that they can slide without excessive friction and without damage.

The advantages and disadvantages of grease as compared with oil follow mainly from the differences in their flow properties. In simple terms it is harder to introduce a grease between two bearing surfaces to form a lubricating film because of its resistance to flow, but once the film is formed it is more difficult to remove.

(a) *Advantages*

The following are examples of situations where the higher resistance of
a grease to deformation or flow is beneficial.

(i) *The stop-start problem*

When an oil-lubricated system is shut down, the oil immediately
begins to drain away from many of the bearing surfaces. The oil is
usually at its hottest before shut-down, and so its viscosity will be
low and it will drain quickly. The result is that when the system is
re-started many of the bearing surfaces will lack lubricant until a
fresh supply of oil has reached them.

Polymer-thickened multigrade oils have some advantage in this
situation, but greases have almost no tendency to drain away
because of their semi-solid consistency. They therefore maintain an
effective film on bearing surfaces during a shut-down.

(ii) *Squeeze film lubrication*

Apart from the problems of lubrication when one bearing surface
slides over another, there are also many situations where one
bearing surface moves towards the other, either with or without any
sideways movement. Under these circumstances lubrication is
assisted by what is called the 'squeeze film effect' (see Chapter 1,
p. 5). The effectiveness of the protection given by an oil film
increases with the viscosity of the oil. A grease is particularly
effective in squeeze film conditions because it acts almost as an
elastic solid. Some examples are in plain thrust bearings, motor
vehicle chassis bearings, and recoil mechanisms.

(iii) *Sealing problems*

Seals are used with bearings for two reasons: (a) to prevent
lubricant from leaking from the bearing and (b) to prevent dirt and
other contaminants from entering. With oil lubrication the ability to
flow readily makes it easy for oil to leave the bearing, and even if
seals are used it is difficult to prevent the oil completely from
leaking. With grease lubrication the poor flow properties prevent
the grease from leaving the bearing easily, so that seals may not be
required, while the surplus grease itself acts as an effective seal
against contaminants. Greases are, in fact, sometimes used to seal
systems containing water or gases, the grease being packed into the
annular space between two rings around a rotating or reciprocating
shaft. The device is known as a packed gland.

(iv) *Supply of surplus lubricant*

The fluidity of oil makes it necessary to use complicated systems
to control or feed an extra supply to the bearing. The semi-solid
nature of grease enables it to be fed into the spaces around the

working surfaces with little or no special design feature being needed to control it.

(v) *Contamination of product*
In the manufacture of critically clean products such as pharmaceuticals, food, and textiles, it is difficult to prevent contamination of the product by oil, which tends to leak or splash. The problem is much more easily controlled with grease.

(vi) *Use of solid additives*
Insoluble solid additives, such as molybdenum disulphide and graphite, remain fully dispersed in grease, whereas they tend in time to settle out of oils.

(b) *Disadvantages*

Some of the disadvantages of greases are also due to the resistance to flow, but others are related to the instability of the gel structure.

(i) *Reduced cooling/heat transfer*
Oil can transfer heat by forced flow or convection. In this way it can be very effective in cooling bearings or gears, removing frictional heat or ambient heat. Grease is almost completely ineffective in heat transfer.

(ii) *Limitations on bearing speed*
Because of its high effective viscosity, grease causes higher viscous friction than oil. As bearing speeds increase, the power absorbed due to viscous friction also increases, and this power is converted to heat. Grease is ineffective as a coolant, hence the temperature rises rapidly to the point where the grease is overheated. This imposes a lower speed limit on grease-lubricated bearings than on similar oil-lubricated bearings.

(iii) *Poorer storage stability*
If an oil is stored in a non-corrodable container without air, its life is almost indefinite. Greases under the same conditions are less stable, and after long storage some may separate, soften, harden, darken, or in the case of soap-thickened greases, become rancid.

(iv) *Lack of uniformity*
Greases tend to be more variable than oils because of the batch manufacturing processes and the poor predictability of the processes.

(v) *Compatibility*
Provided that lubricating oils are of the same type (e.g., mineral, ester, silicone) they rarely present any problems of compatibility. On the other hand, greases may not be fully compatible with each other even if they use similar base oils and thickeners.

(vi) *Lower resistance to oxidation*
     The effect of oxidation products on the gel stability makes greases
     more susceptible to oxidation problems than oils.

7.7 Selection and applications of greases

The selection of a grease for a particular application depends on three
factors: the type of component, the temperature, and the environment. In
practice a wide variety of applications can be satisfactorily handled by a
so-called 'multi-purpose grease'. This is usually an NLGI No. 2 mineral
oil grease, and is often lithium hydroxystearate or aluminium complex-
thickened.

Where the manufacturer of the equipment recommends a particular
type of grease, or even a particular brand, this recommendation should
generally be followed unless there are good reasons for not doing so.

The recommendation may include the type of base oil, type of
thickener, consistency, base oil viscosity, and any requirement for EP
additives.

Where no specific recommendations exist, the following guidelines
can be used.

(a) *Type of component*

(i)   *Rolling bearings*
Greases for use in rolling bearings will generally be of NLGI Grades
Nos 1, 2, 3, or 4. A No. 2 would normally be the first choice. The
factors involved in deciding which grade to use are as follows.

Speed   The higher the speed the harder the grease, so for high speeds
use No. 3. The maximum outer race speed for grease-lubricated rolling
bearings is about 15 m/s (3000 ft/min).
Bearing size   Use a harder grease in a large bearing – No. 3 or even
No. 4.
Sealing   If there is a particular need for effective sealing, use a harder
grease – No. 3 or No. 4.
Shock loads or vibration   Shock load or vibration tends to cause grease
to slump into the moving components causing churning. A harder grease
reduces the problem, so use No 3 or No 4. (Note that normal steady
loading does not strongly influence the grade.)
Temperature   For starting in very cold climates a softer grade may be
helpful. The influence of temperatures is considered in greater detail
later.
Centralized grease supply   Use a softer grade – No. 2, No. 1 or even
Number 0.

Thus a small low-speed bearing in a cold environment might require a No. 1 grease, while a large bearing with shock loads or a particular sealing problem might use a No. 3, but if in doubt use a No. 2.

The quantity of grease used is quite critical. If the bearing and housing are completely filled, the surplus grease cannot escape from the working zone and will be continually churned, causing over-heating. Various recommendations range from 25–60 percent of the total volume for ball bearings and about 30 percent for roller bearings. For ball bearings the quantity depends on the speed, and 25 percent is probably best for very high-speed bearings and 40–50 percent for slow ones. For extremely slow bearings there is much less tendency to over-heat, and even full packing may not cause any serious problem, except for unnecessarily high viscous drag.

If the temperature of the bearing can be monitored, it will be found that in a properly filled bearing the temperature first rises, to as much as 70 °C–80 °C, and then falls to a steady running temperature of 30 °C–40 °C. If there is too much grease, the temperature will rise and remain high, whereas with insufficient grease the initial rise will not occur, or will be reduced.

Whatever percentage grease fill is used, it is important to make sure that the races and rolling elements are fully coated. Due to the poor flow properties of grease, an uncoated area may remain dry for some time and become damaged. It is therefore good practice when assembling a bearing to apply a thin film of grease over all the running surfaces.

With a few types of roller bearing where sliding takes place under load, such as tapered roller bearings, an EP grease should be used.

This problem of the optimum percentage packing raises difficulties in re-lubrication, whether by grease gun or centralized system. Unless grease valves are fitted to allow the surplus to escape, the greatest danger is that too much will be added. Bearing manufacturers sometimes give guidance to the frequency and amount of re-lubrication. If the initial grease pack is correct, the bearing does not overheat, and there are no obvious signs of leakage, it may not be necessary to add grease until the time comes to strip and re-lubricate the bearing. If there are signs of leakage, then the amount of leakage may indicate the amount of make-up required, but it will normally be advisable to add only a small proportion of the total fill, perhaps 5 percent at a time.

(ii)   *Plain bearings*

Although grease may be the first choice to be considered for rolling bearings, this is probably not true for plain bearings.

However, for plain journal bearings or sleeve bearings, greases may be the best choice when the speed of rotation is slow, the clearances are large, or the bearing is subject to shock loading or frequent stops, or is inaccessible for re-lubrication.

The limiting speed of rotation for grease lubrication of plain bearings is much lower than for rolling bearings of comparable size, as can be seen in Fig. 2.1. This is because in a plain bearing a higher quantity of grease will be continuously sheared so that more heat will be generated. The actual speed limit is about 1–2 m/s (200–400 ft/min), representing 400 rpm in a 5 cm diameter bearing or 50 rpm in an 80 cm diameter bearing.

Many plain journal bearings can be satisfactorily lubricated for quite long periods with only a small initial charge of grease. However, it must be remembered that in such a bearing the whole of the charge is being worked all the time. It follows that, however stable the grease, there will come a time when it softens too much to remain in the bearing, or oxidizes, darkens, and hardens or dries. The time this takes depends on the speed, the clearance, the temperature, and the stability of the grease.

Suitable greases are of No. 2 or No. 1 consistency, not tacky and not showing significant oil separation, but softening rather than hardening on prolonged working. The initial charge should be sufficient to ensure that the whole of the bearing and journal surfaces are coated and some grease is forced axially out of the bearing.

Preferably, some means of re-greasing should be provided, either hand-operated grease gun, automatic pump, or centralized system. The grease can be fed into the bearing through a central hole associated with an axial groove and located between 90 degrees and 180 degrees from the loaded zone on the entry side. The groove should extend to about 80 percent of the full axial length of the bearing, and should be wider and deeper than for oil lubrication to allow for the higher viscosity of grease.

Wherever possible, fresh grease should be fed into the bearing until all the old grease has been displaced and the fresh grease exudes from the bearing. The quantity and quality of the old grease indicates whether a shorter re-greasing period is required.

Certain other types of plain bearing, such as sliders or thrust washers, may be grease-lubricated provided that the geometry ensures that grease remains present. Tilting pad thrust or journal bearings cannot satisfactorily be lubricated with grease.

(iii) *Enclosed gears*

Grease may be used to lubricate gear sets where these are small, slow, intermittent, or have a tendency to leak oil.

The main problem in grease lubrication of gears is again that of poor heat transfer. Small boxes have a higher specific surface area (i.e., surface area divided by weight or volume) and therefore lose heat more easily. Slow or intermittent gears have less tendency to generate heat.

For intermittent operation a major advantage of grease is that a film of lubricant is maintained on the moving parts while the box is not operating. Adequate lubrication is therefore obtained immediately on starting. The grease should be a soft one, No. 1, 0 or even 00, to give low viscous friction.

From two to five percent of molybdenum disulphide may be included in the grease, especially if the gears are highly loaded. This improves the maintenance of a low-friction film on the surfaces while stationary, and improves the load-carrying capacity.

For small gearboxes which run more continuously the grease should again be a relatively soft one, No. 0 or 1. It should not soften too much with continuous working, or stiffen quickly when movement stops. Greases which soften quickly when sheared but stiffen quickly when shearing stops are called thixotropic.

The larger the gearbox, the harder should be the grease. If a soft grease is used in large quantity, it does not remain clear of the moving parts, but slumps into them. As a result it is continually worked or churned, leading to over-heating and breakdown of the grease.

Thixotropic greases are particularly useful where grease is being used to reduce leakage. The grease in contact with the gear surfaces is soft while being worked, but elsewhere in the gearbox it is stiffer and does not leak. For many purposes a thixotropic No. 1, 0, or 00 grease is adequate. For slow, heavily loaded gears a No. 2 grease with slight thixotropic properties may be better.

A small quantity (2–4 percent) of molybdenum disulphide may again be used to improve load-carrying capacity and prevent scuffing at high loads. Where a molybdenum disulphide-containing grease is recommended, the user should buy one rather than adding molybdenum disulphide to an existing grease, as this can affect the structure and stability.

(iv) *Open gears*

The advantage of grease for open gears is that it is more effectively retained on the gear teeth than an oil, and tacky or adhesive additives such as bitumen are often used in open gear greases, to provide even better adhesion. The grease may be No. 0, 1, or 2, depending on the size, speed and load, a soft grease being used for large, slow or lightly-loaded gears. For light or intermittent duty the initial grease pack may last for a long time, but for large gears which are continuously operating a mechanical feed or even a grease spray may be used. The speed limit for grease-lubricated open gears is similar to that for plain bearings, but certainly not more than 5 m/s (1000 ft/min).

(v)   *Other components*
The most widely used grease for lubricating most flexible couplings is a
No. 2 multipurpose grease containing an anti-oxidant. The same is true
of a wide range of other components.

(b)   *Temperature*

The general effects of temperature on greases were discussed in Section
6.5 (b). The following gives some advice on grease selection for specific
temperature situations.

(i)   *Low temperatures*
The most important factor in determining the low temperature limit for
operation of a grease is the viscosity or pour point of the base oil. Table
7.4 lists the low temperature limits for several different types of grease.
   The consistency is also important, particularly when the equipment
must be started at low temperature. It is, however, the consistency at the
low temperature which is important, not at the usual temperature of
25 °C which is used for laboratory testing. Since the consistency is not
normally measured, and not easily measured, at low temperature, the
best guidance is to select the base oil type first and then select a suitable
grade made with that base oil. For example, a di-ester grease may be
suitable for use at −50 °C and a No. 2 will probably give easier starting
at −50 °C than a No. 3.

(ii)   *High temperatures*
The maximum temperature for use of a grease depends on both the drop
point and the stability of the base oil and other components. These
effects were summarized in Section 7.5 (b).
   A grease may still give effective lubrication above its drop point, but

Table 7.4   Low temperature limits for various greases

Base oil	Thickener	Minimum temperature (°C)
Mineral oil	Calcium soap	−20
Mineral oil	Sodium soap	0
Mineral oil	Lithium soap	−40
Mineral oil	Bentonite clay	−30
Synthetic hydrocarbon	Lithium or aluminium complex	−70
Diester	Lithium soap	−75
Diester	Bentonite clay	−55
Silicone	Lithium soap	−55
Silicone	Dye	−75

it will no longer be a grease. For example, a sodium soap grease might be liquid at 130 °C, but provided the base oil is viscous enough, the liquid could still give satisfactory lubrication, although the particular advantages of a grease would be lost.

A grease cannot usually give satisfactory lubrication once its base oil has decomposed, due to excessive temperature. However, the rate of decomposition depends on temperature and time, as shown in Fig. 7.6. It follows that greases may be usable for short periods at temperatures higher than those at which they start to decompose.

(c) *Environmental factors*

The types of environmental factor which affect grease selection include: vacuum (in spacecraft, aircraft, or vacuum equipment), dirty environments (such as mining and quarrying), clean environments (such as food, textile and pharmaceutical industries), and high fire-risk situations (such as coal mining, aviation and certain chemical processes). The following examples will illustrate the way in which these factors are taken into account in grease selection.

(i) *Coal mining*
Greases are widely used in coal mining because of their ability to seal out dirt. They are also preferred to oil because they reduce leakage and because their poor flow reduces fire risks. They therefore tend to be fairly stiff, either No. 2 or No. 3, to give maximum sealing and minimum leakage. Where softer greases are desirable, as in enclosed gearboxes, they should be thixotropic, as explained previously.

Although greases present lower fire risk than oils, they can nevertheless initiate fires. A number of underground fires have been caused by grease-lubricated conveyor bearings, especially when over-greasing has resulted in accumulations of grease around the bearing locations. There is therefore some interest in fire-resistant greases. Those which have been developed include some using phosphate ester base oils.

(ii) *Food Industry*
The dominant problem in the food industry, as in the pharmaceutical and textile industries, is to prevent contamination of the products. Greases are again used in order to cut down the risk of leakage, and the tendency is to use No. 2 or No. 3. In addition, many governments restrict the types of thickener and base oil which may be used, so that any accidental contamination will not be harmful. Table 7.5 lists some of the components which are permitted in the United States for use in greases for food-making machinery.

Petrolatum (petroleum jelly) is also used as a lubricant for certain low-

Table 7.5   Some substances permitted in greases for the food industry

Base oils	Thickeners	Additives
1. White oils	1. Silica	1. Phenylnaphthylamine
2. Polybutene	2. Polyisobutylene	2. Butylhydroxyanisole
3. Methyl silicone	3. Polyethylene	3. Sodium nitrite
4. Castor oil		4. Isopropyl oleate

duty applications in the food industry. However, its mediocre load-carrying capacity and temperature limitations restrict its use.

(iii) *Aviation*

Greases are used instead of oil wherever possible in aircraft because of the weight penalty involved in providing oil feed systems and sealing. In general, oil is used only where there is a serious cooling requirement, as in the engines, or where the oil serves some other primary purpose, such as in hydraulic systems, brake systems, or undercarriage oleo assemblies.

Although many thousands of aircraft in general aviation have no need to operate at extremely low temperatures, the development of such aircraft has been dominated by the North American market, where the surface temperature in winter may reach −45 °C. In addition, all modern airline and military aircraft are required to operate at high altitudes and high latitudes, where temperatures may be even lower – reaching −75 °C in places.

Modern high-speed aircraft also experience high temperatures due to aerodynamic heating. For many applications greases are required to operate at 175 °C, while for a few specialized applications the specified temperature is over 200 °C.

Thus a supersonic aircraft may be required to start up after standing at −50 °C, to take-off almost immediately, and to be experiencing component temperatures over 150 °C a short time later when flying at supersonic speeds. The greases must therefore be capable of operating satisfactorily from below −50 °C to above 150 °C.

Fortunately certain base oils are capable of covering such a wide temperature range. Clay-thickened diester greases have been available for many years which can be used from −54 °C to 150 °C. More recently, synthetic hydrocarbons thickened with modified clays have been developed to cover even wider temperature ranges.

(iv) *Spacecraft*

In the early days of space exploration it was assumed that lubricants would have to have exceptionally low volatility to withstand the high vacuum of space. Certain greases used in early spacecraft were based on chlorinated silicones or special low-volatility mineral oils, but these were not very good lubricants.

More recently it has been found that the evaporation of oil from grease-lubricated rolling bearings can be controlled by the use of labyrinth seals, and more conventional mineral-based greases are now sometimes used in such application.

7.8 Methods of applying greases

Where grease is used as the initial fill for rolling bearings and such components as doorlocks, car window winders, and so on, it is very often applied by hand. For filling large bearings or gearboxes, 'paddles' shaped like small table-tennis bats may be used.

To increase the supply of grease to the component in service, a reservoir may be supplied in the form of a cup or block of grease. The greases for such systems are stiffer than would otherwise be used – No. 3 or 4 for a 'cup grease' and No. 5 or 6 for a 'block grease'.

By far the most common method of re-greasing a mechanism is by means of a grease gun, applied through a nipple. Figure 7.7 shows a typical grease gun.

One of the problems in re-greasing is to ensure that the right quantity is supplied. In general, the tendency is to supply too much. With simple bushes such as many of those in motor vehicles, any surplus will easily escape from the other end of the bush or through the rubber seals. With more complex components it may be desirable to remove a drain plug during re-greasing and again after a few hours' operation to allow any surplus to escape.

A modern development is a grease valve, which allows any surplus to escape under pressure, but prevents dirt from entering.

Grease can also be supplied automatically by means of a grease feed system, similar in many ways to those used for oil. Greases used with

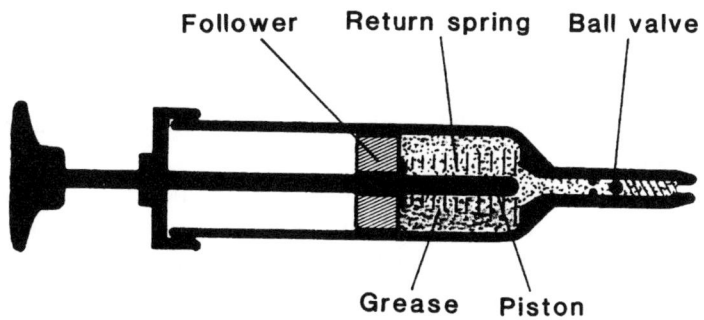

Figure 7.7   Cross-section of a typical grease-gun

centralized systems should preferably be soft – No. 0 or 1 – but No. 2 greases are sometimes used.

All centralized grease lubrication systems are total loss systems, and the various types are similar to those described in Section 5.2 for total loss oil systems. Because of the poor flow properties of grease, high pressures can be generated, so that centralized systems tend to be very robust. Pipe sizes must generally be larger, and pipe runs short to avoid plugging and percolation of base oil through a plug of thickener.

7.9 Anti-seize and anti-scuffing compounds

These materials are perhaps best defined in terms of the job they are designed to do. The main purpose of anti-seize compounds is to prevent two metal surfaces from binding together or actually seizing under conditions of high load and either very low speed or stationary contact. For example, an anti-seize compound might be applied to the threads of a nut and bolt to prevent them from seizing while being tightened or to prevent them from binding together while stationary after having been tightened.

(The terms 'binding' and 'seizing' do not have any widely recognized definitions. Here 'binding' is used to describe the very high friction which can arise when two surfaces are loaded together, but which disappears if the load is removed without leaving any damage. 'Seizing' is used to describe the high friction sticking of two surfaces under load which results in some surface damage and often persists after the load is removed.)

Anti-scuffing compounds are designed to prevent damage to surfaces while sliding against each other at high load and low speed. Scuffing is often the first stage in seizure of two components, so that the requirements of an anti-scuffing compound are similar to those of an anti-seize compound. The same material can often be used for both purposes.

The general characteristics of both anti-seize and anti-scuffing compounds are:

(a) high viscosity, to prevent them from being forced out easily;
(b) low friction to enable the surfaces to move without damage under high load;
(c) high load-carrying capacity, to prevent asperities on the two surfaces from welding together when sliding under high load;
(d) good physical and chemical stability, and inertness to the contacting surfaces, because they may be expected to prevent seizure of components bolted or clamped together for very long periods.

Ordinary greases and high viscosity oils may be used as anti-seize and anti-scuffing compounds. However, the purpose-made compounds usually consist of high concentrations of solid lubricants in grease, oil, or petrolatum. Some examples of typical anti-seize and anti-scuffing compounds are given below.

(a) Between 40 and 70 percent of molybdenum disulphide powder in a grease or petrolatum (soft petroleum wax, petroleum jelly). This acts as an anti-scuffing paste or anti-seize at low temperatures and as an anti-seize up to over 250 °C. At the high temperatures the grease or petrolatum decomposes, but the concentration of molybdenum disulphide in the residue is high enough to maintain effective lubrication.

(b) Fifty percent of graphite in a base of petrolatum or polyglycol. This is usable on threaded fittings such as sparking plugs up to 500 °C. At high temperatures the base material decomposes, leaving little (with petrolatum) or no (with polyglycol) residue, and the graphite remains as an effective anti-seize until it too begins to oxidize above 500 °C. For electrical components such as sparking plugs the graphite also improves electrical conductivity through the threaded joint.

(c) Graphite in the form of a paste in a volatile solvent such as ethyl alcohol or acetone. This is a technique for applying a thick film of dry graphite powder to a metal surface, from which the volatile solvent rapidly evaporates.

(d) A paste of graphite in water, or in a non-flammable fluid such as a fluorocarbon, is used an an anti-seize for low-pressure oxygen systems.

(e) Pastes of low-friction metal powders such as lead and copper in petrolatum or polyglycol, or similar powders in a low-melting polymer or flux. These metal powders are effective anti-seizes to temperatures as high as 1000 °C, and are sometimes used on threads of sparking plugs, where the metal powders help to ensure electrical continuity.

It is very important to remember that anti-seize compounds containing a high concentration of molybdenum disulphide will generate very low friction during assembly. Where an assembly torque is specified for a bolted assembly, the resulting bolt tension could be as much as ten times as high if a molybdenum disulphide anti-seize is used. This has been known to cause tensile failure of bolts during assembly or rapid fatigue failure in service.

Chapter 8

Dry Bearings and Solid Lubrication

8.1 Mechanism of solid lubrication

Solid lubrication is simply the lubrication of two surfaces in moving contact by means of solid materials interposed between them.

Depending on the nature of the two surfaces, a wide variety of solid materials can reduce friction and prevent seizure. For example, dust, sand, or gravel on the surface of a road can cause vehicles to skid because they decrease friction between tyres and the road surface. Technically speaking, they are acting as lubricants, but of course, no one would normally recommend dust, sand, or gravel as solid lubricants.

Hundreds of different solids have in fact been tested for use as solid lubricants, and probably scores of them have been used in practical applications. They include materials such as mica, talc, and chalk. However, the ability to reduce friction is only one of the important properties of solid lubricants, and most materials have serious faults which rule them out as effective lubricants. As a result the vast majority of solid lubricant applications are met by only three materials: graphite, molybdenum disulphide, and PTFE (polytetrafluoroethylene).

The properties which are desirable for a good solid lubricant are:

(a) low but constant and controlled friction;
(b) chemically stable over the required temperature range;
(c) no attack or damage to the bearing materials;
(d) preferably adhere strongly to one or both bearing surfaces so that it is not rapidly lost from the bearing;
(e) sufficient resistance to wear to provide a useful life;
(f) easy to apply in a controlled manner;
(g) non-toxic;
(h) economical.

These requirements are very demanding, and most of the successful compounds are either polymers, inorganic powders, or metals. Some of these are listed in Table 8.1. Many of them are incorporated in composites. These are mixed materials in which, for example, a soft bulk solid may be reinforced to make it stronger, or a lubricant powder may be dispersed in a strong solid to reduce its friction, or a binder may be used to hold together the particles of a powdered lubricant. The variety of solid lubricating composites is enormous, but only a few types have been used at all widely; the important types will be described later.

Graphite and PTFE, and many other solid lubricants and composites can be used to construct bearings or bearing components. These are often known as 'dry bearings', but the expression is not entirely satisfactory. Under certain circumstances more conventional types of bearing can be operated without lubricants, for example at light loads, low speeds, or very low temperatures. They would also be called 'dry bearings'.

The term 'dry-bearing materials' is more acceptable, since it means materials which are specifically used to manufacture bearings intended to operate without lubricants. In practice solid lubricants and dry bearing materials are very similar and may even be identical. They are best considered as a single group of materials.

The three important solid lubricants, and many of the others, all have a layered structure, as shown in Fig. 8.1.

Table 8.1   Some materials used as solid lubricants

Layer-lattice compounds

Molybdenum disulphide	Graphite
Tungsten diselenide	Tungsten disulphide
Niobium diselenide	Tantalum disulphide
Calcium fluoride	Graphite fluoride

Polymers

PTFE	Nylon
PTFCE	Acetal
PVF_2	Polyimide
FEP	Polyphenylene sulphide
PEEK	

Metals

Lead	Tin
Gold	Silver
Indium	

Other inorganics

Molybdic oxide	Boron trioxide
Lead monoxide	Boron nitride

Figure 8.1   Layered structure of many solid lubricants

With graphite, molybdenum disulphide, calcium fluoride, and others, the layers consist of flat sheets of atoms or molecules, and the structure is called a 'layer-lattice' structure. With PTFE and some other polymers the layers consist of long parallel straight molecular chains.

The important effect is the same in all these cases, namely that the materials can shear more easily parallel to the layers than across them. As a result they can support relatively heavy loads at right angles to the layers while still being able to slide easily parallel to the layers. The theoretical analysis is given in the Appendix to this chapter, but briefly the coefficient of friction is equal to the shear stress parallel to the layers divided by the yield stress or hardness perpendicular to the layers.

With PTFE, and with molybdenum disulphide and some similar materials, the shear stress parallel to the layers is inherently low. In other words, the low shear stress is a property of the material itself. With certain other layer-lattice materials, particularly graphite, the shear stress parallel to the layers is not particularly low when the material is pure or clean, but becomes low when contaminants like water vapour are present.

Since low friction only occurs parallel to the layers, it follows that these solid lubricants will only be effective when their layers are parallel to the direction of sliding.

It is also important that the solid lubricant should adhere strongly to the bearing surface, otherwise it would be easily rubbed away and give very short service life. The best of the solid lubricants all stick very strongly to metal surfaces or even to many non-metals. Some layer-lattice compounds, including talc and mica, will not adhere, and are therefore of limited value for lubrication.

A useful characteristic for indicating the lubricating capacity of a solid lubricant is the PV factor. This is the highest product of pressure and velocity at which the lubricant can be used satisfactorily. It is usually quoted as

$kN/m^2 \times m/s$ or kN/ms

but may still be quoted in Imperial Units as

$psi \times ft/min$

The PV factor represents a convenient over-simplification; it masks the fact that for one material the limiting PV may be dominated by the load, and for another by the speed. In other words, two materials may have the same PV limit, but one can be used at higher load than the other, while the second can be used at higher speed. For particular applications it may sometimes be important to assess the load and speed limits separately.

8.2 Advantages and disadvantages of solid lubricants

Compared with oils and greases, solid lubricants are far less widely used. Oils and greases might be considered as the 'normal' lubricants, which would be used unless there were good reasons for not doing so. Solid lubricants, on the other hand, are used where unusual circumstances exist which make oils and greases unsuitable.

It was pointed out in Chapter 6 in connection with greases that their advantages and disadvantages stem from their inferior flow properties. In this respect solid lubricants represent an even more extreme case, since they have virtually no flow properties. Their advantages and disadvantages result mainly from this lack of flow.

(a) *Advantages*
   (i) Practically no tendency to flow, creep or migrate, so that they can be relied on to remain in place for long periods. This can be particularly important for intermittent use or where equipment must operate quickly and satisfactorily after long storage.
  (ii) Minimum tendency to contaminate products or environment.
 (iii) Solid form plus high structural strength of some solid lubricants enables bearing components to be manufactured from them.
  (iv) Very low volatility enables them to be used in high vacuum.
   (v) Often usable at very high or very low temperatures.
  (vi) Chemical inertness enables some of them to be used in reactive chemical environments.
 (vii) Generally stable to radioactivity.
(viii) Can maintain lubricant film under high load, low speed, or vibration.

(b) *Disadvantages*
   (i) Difficult and often impossible to feed or replenish.
  (ii) Inevitably wear away in use so that their service life (as opposed to storage life) is limited.
 (iii) Generally poor thermal conductivity limits their maximum sliding speed.

(iv) Thermal expansion very different from metals, so that loss of clearance can occur with temperature changes when they are used in conjunction with metals.

These advantages and disadvantages apply to different extents to different solid lubricants.

8.3 Graphite

Graphite was almost certainly the first solid lubricant to be used on a large scale, and its use goes back to prehistoric times. It is a grey-black crystalline form of carbon in which the carbon atoms are arranged hexagonally in regular layers, as shown in Fig. 8.2.

In this hexagonal structure the bonds between the carbon atoms within a layer are strong chemical (covalent) bonds, so that the layers are strong and the crystals strongly resist bending or breaking of the layers. The bonds between the layers are weak (van der Waals) forces, so that the crystals can be made to split easily between layers, and the layers will slide readily one over the other.

Graphite occurs naturally in large veins or in flakes. These natural graphites vary enormously in their degree of crystallinity and in their purity. The purity varies from 80 percent to 90 percent but can be increased in refining to as high as 98 percent.

Graphite can also be manufactured by a process known as graphitizing, the greater part being made from petroleum coke by heating to 2600°–3000 °C. Synthetic graphite is highly crystalline and very pure, containing on average 98.5 percent of carbon.

Although all graphites – natural and synthetic – contain a very high

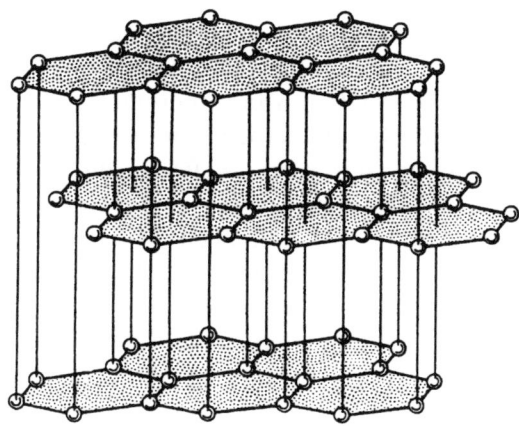

Figure 8.2   Hexagonal lattice structure of graphite

proportion of carbon, some of this carbon may be non-graphitic in form, and the frictional behaviour varies with the graphite content. They also differ considerably in the degree of crystallization, uniformity or randomness of their crystal orientation, and their lubricating and other properties also differ as a result of this variability. The general non-specialist user is therefore strongly recommended to buy graphite in the form of suitable finished branded products and leave the manufacturer to worry about the great variability of the raw materials.

The layer-lattice structure of graphite would appear to give good natural low-friction properties. In fact, graphite only gives low friction when it is contaminated or 'intercalated' by water vapour or other condensable vapours. During the Second World War, when aircraft started to operate regularly at high altitudes, it was found that their electrical brushes, made of carbon-graphite, wore very rapidly – so rapidly in fact that the process was called 'dusting'. It was found to be due to the removal of adsorbed water vapour at the low pressures at high altitude.

The problem can be overcome by introducing a less volatile vapour such as bromo-pentane, but this will also be removed by lengthy use in vacuum. Unless good electrical conductivity is needed, it is usually better to avoid the use of graphite in vacuum or dry atmospheres. Where the electrical conductivity is needed, high-density graphite will work satisfactorily for longer periods than porous graphite, presumably because it takes longer to remove the absorbed vapours. For very long use some molybdenum disulphide can be incorporated to maintain low friction.

High-quality graphite with water vapour present will give coefficients of friction varying from 0.05 at high contact pressures to 0.15 at low pressures; the friction is smooth and steady. Low friction can be maintained at temperatures up to about 200 °C if the continued presence of water vapour or other suitable contaminants can be ensured. In practice this is usually difficult, and above 160 °C in air the coefficient of friction tends to rise to 0.3–0.4. More important in many applications is the fact that the friction becomes much more irregular, and there is a tendency to rough sliding and wear. A few additives such as nickel chloride have been shown to improve friction for limited periods in this temperature range, but they have not been widely used. For practical purposes there is a temperature range from 200 °C to 350° in which graphite is largely ineffective as a lubricant. However, this is also a temperature range for which few satisfactory lubricants are available, and in spite of its shortcomings graphite may be better than no lubricant at all.

Above 350 °C graphite again provides low friction, possibly because of some reaction with surface oxide on metal surfaces. The friction again starts to increase rapidly at about 540 °C when the graphite starts to oxidize in air. As a result graphite can be used to lubricate many hot industrial metalworking processes, such as wire-drawing, tube-drawing, and forging. It can be used for short periods at even higher temperatures because it leaves little or no harmful residue when it oxidizes, but with some metals, including steels, it may cause carburization and embrittlement.

Graphite adheres readily and firmly to metal surfaces and will embed into softer materials such as polymers. The edges of the crystal planes are quite sharp and hard, and therefore abrasive. It is important for the crystals to be oriented parallel to bearing surfaces in order to reduce the abrasiveness and give the lowest possible friction. Because of the easier shear parallel to the crystal planes graphite tends to form flat particles which preferentially stick to flat surfaces in the desired orientation.

If graphite is ground to progressively finer particle size, the proportion of crystal edges exposed becomes higher, and the friction and abrasion increase. There is also an apparent tendency with very fine grinding for the low-friction hexagonal crystal structure to change to a higher friction 'turbostratic' form when the average particle size is less than about 0.1 μm (4 μin). The preferred particle size range is probably from about 1 μm (40 μin) to 40 μm (1.6 thou in).

Graphite will adhere readily to surfaces either from a solid block (as in writing with a pencil) or from dispersions in liquids. It can also be made to adhere more strongly by means of adhesive binders, usually epoxy, phenolic or alkyd, which may also be included in a dispersion. A wide variety of graphite dispersions are commercially available. Table 8.2 lists a few of the main types and their applications.

Table 8.2 Some graphite dispersions

Liquid	Graphite concentration (%)	Applications
Water	20–30	Mould lubricant (release agent); tool lubricant; rubber lubricant; electrically conducting coating
Mineral oil	10	Mould and tool lubricant
	35–40	Metalforming lubricant; anti-seize
Castor oil	10	Mould lubricant for natural rubber
Isopropyl alcohol	10–20	Dry film mould lubricant, anti-seize and electrically conducting coating
White spirit	50	Anti-seize; release agent; high-temperature lubricant
Polyglycol	10–20	Extreme temperature lubricant

As a powder or film, graphite can support loads of the order of 70 MN/m^2 (10 000 psi) depending on the amount of powder or the thickness of the film. As with most solid lubricants the maximum sliding speed is limited. However, because graphite has a higher thermal conductivity it can be used at higher speeds than molybdenum disulphide or polymers.

The maximum PV for graphite powder, rubbed film, bonded film or solid block is probably about 0.7 MN/ms (20 000 psi × ft/min). If a lubricating oil is used with a film or a solid block of carbon/graphite, the PV limit is probably ten times as high, because of the cooling effect of the oil.

Disadvantages of graphite or graphitized carbon in block form are structural weakness, porosity, and difficulty of accurate machining. These can be improved by impregnating the carbon with a resin or metal, but the friction then rises to 0.10–0.25, and the maximum PV is about 0.07 MN/ms (2000 psi × ft/min). It is also reported to encourage corrosion in a sea-water environment.

The main properties of graphite as a lubricant can, therefore, be summed up as follows:

- low friction 0.05–0.15 depending on pressure;
- maximum PV 0.7 MN/ms dry;
- good adhesion;
- good thermal and electrical conductivity;
- usable at low temperatures and to 540°C in air;
- poor performance in vacuum or when very dry;
- good performance in presence of liquids;
- very complex and variable materials;
- black and, therefore, unacceptable for certain processes.

8.4 Molybdenum disulphide and similar compounds

Molybdenum disulphide has also been used for several centuries, although its early history is not clear because it was confused with graphite. It was certainly used by the California gold-miners ('Forty-niners') in 1848–50 to lubricate the axles of their wagons.

It is found naturally as the ore molybdenite, the main source of molybdenum. This is a crude hexagonal molybdenum disulphide, and is found in large masses and in immense quantities in some parts of the world.

The chemical formula is MoS_2, with one molybdenum atom joined to two sulphur atoms. Like graphite, it has a hexagonal layer-lattice structure, as well as other, non-lubricating, forms. The structure consists of alternate layers of molybdenum and sulphur atoms.

Each layer of molybdenum atoms is sandwiched between two layers of sulphur atoms. The bonds between molybdenum and sulphur layers are strong covalent chemical bonds. Each triple layer or sandwich of molybdenum disulphide molecules is joined to the next layer only by weak 'van der Waals' forces between the layers of sulphur atoms, so that the crystal can easily split along the cleavage faces between the molybdenum disulphide layers. We therefore have again the low friction characteristic of a material which is strong in one direction (perpendicular to the layers) but easily sheared in another direction (parallel to layers).

There is in fact a family of chemical compounds similar to molybdenum disulphide, known by the cumbersome name of 'lubricating dichalcogenides'. They include the disulphides, diselenides, and ditellurides of molybdenum, tantalum, tungsten, and niobium. Of these only molybdenum disulphide occurs naturally. All the others have to be made synthetically, and are, therefore, expensive.

The advantages of the other compounds over molybdenum disulphide are very limited. Tungsten disulphide has a higher oxidation temperature, at 500 °C in air, and both it and tungsten diselenide oxidize more slowly than molybdenum disulphide. Niobium diselenide has been used to some extent in electrical brushes for high vacuum because of its better electrical conductivity. In fact, composites containing molybdenum disulphide have proved to be at least as good. As a result there is now very little interest in these compounds, and only molybdenum disulphide is of major importance.

Crude natural molybdenite contains high concentrations of impurities. These are reduced in refining to less than 2 percent, and much of this is harmless carbon from oil used in the purification. A small amount of the remainder is silica, and in some grades this is removed by treatment with hydrofluoric acid, but it is doubtful if this treatment is justified. As with graphite, the edges of the small crystallites of molybdenum disulphide are abrasive, so the removal of silica gives little, if any, improvement.

Molybdenum disulphide differs from graphite mainly in that its low friction is an inherent property and does not depend on the presence of absorbed vapours. Because of this, it can be used satisfactorily in high vacuum, and it has been used for many applications in spacecraft (see Fig. 8.3).

It adheres even more strongly than graphite to metal and other surfaces. A useful film can be obtained by simply rubbing or 'burnishing' molybdenum disulphide powder onto a metal surface with cottonwool or cloth. The rubbing should be done smoothly and

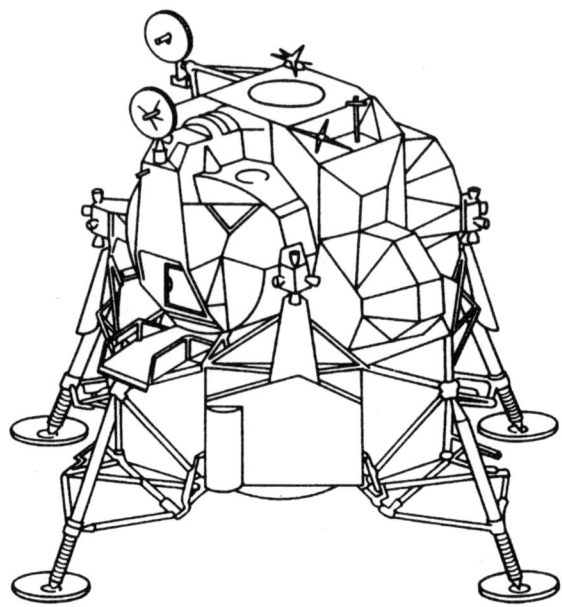

Figure 8.3 Extendible legs on Apollo lunar module lubricated with bonded molybdenum disulphide (courtesy American Metal Climax Inc.)

uniformly, always in the same direction, and will give a glossy blue-black film about 0.1 μm (4 μin) thick. Thicker films, up to 10 μm (0.4 thou in) thick, can be produced by repeated addition of powder and burnishing, and can give lives of many hours at 1000 psi contact pressure at low sliding speed.

Films can also be produced by dipping or spraying with a dispersion of molybdenum disulphide in a volatile solvent or water and allowing the liquid to evaporate. Such a film can at first be easily scraped off, but when rubbed on or burnished it becomes very durable.

Very thin, strong, durable films are produced by a technique called sputtering. Only a few experienced suppliers of such films are available. Much more readily available are bonded films in which a resin such as epoxy, phenolic, silicone or various inorganic compounds is present. The mixture of molybdenum disulphide and resin is dispersed in a solvent and applied by spraying, often from an aerosol can, or by brushing or dipping. The solvent evaporates and the binder is then cured either at room temperature or at temperatures up to 200 °C to give a hard varnish-like film which has very good life. One disadvantage that bonded films have over sputtered films is that they are very much thicker. This means that they are more difficult to incorporate in precision bearing systems.

A great variety of different bonded films has been produced. For high temperatures, inorganic-bonded films should be used, and a silicate-bonded film containing molybdenum disulphide and graphite has been successfully used in many different applications. Unfortunately, however, the presence of the graphite has been shown to encourage corrosion in humid environments, so that its use is now less popular. Other inorganic films are, however, commercially available.

Table 8.3 lists the characteristics of different molybdenum disulphide films.

All of these types of film give their best performance if the metal surface has a finish of 0.5–1 μm CLA (20–40 μin), preferably produced by grit-blasting, and if the surface is then given one of the following pretreatments:

Steel	Phosphate or sulphide
Aluminium	Anodize
Magnesium	Dichromate
Titanium	Anodize or phosphate-fluoride

The coefficient of friction of such films varies from 0.03 at high contact pressure to 0.1 at low pressure.

As with graphite, the proportion of crystal edges present increases if the powder is ground too fine; for most purposes an average particle size of 2–3 μm is probably best. The film should preferably be run-in at low load before full service load is applied, as the running-in process helps to make the film more compact, and therefore stronger.

Films do not adhere so well in the presence of a liquid. Nevertheless dispersions of molybdenum disulphide in lubricating oils and greases are widely used. There is a great deal of argument about the value of molybdenum disulphide in vehicle engine oils, and some early experiments indicated that there was slightly more wear when it was used. More recently fleet tests in large trucks did not report any disadvantages, but did show a small saving in fuel, presumably because of decreased friction.

At concentrations of 3–5 percent in greases, a considerable increase in

Table 8.3 Characteristics of molybdenum disulphide films

Type of film	Film thickness (μm)	Film characteristics
Burnished	0.1–10.0	Very low friction, thin, durable
Sputtered	0.2–2.0	Fairly low friction, very thin, durable
Organic bonded	2.0–40.0	Low friction, thick, high wear rate
Inorganic bonded	3.0–40.0	Low friction, thick, wide temperature range

load-carrying capacity is obtained, and for very highly loaded bearings or moving contacts such a grease may be useful. A further advantage is that if, for some reason, the grease is lost from the bearing, a residual film of molybdenum disulphide on the surfaces can continue to give satisfactory lubrication for a brief period. Greases containing 1 percent of molybdenum disulphide have been used for vehicle wheel bearings, while greases containing 1–3 percent are used for chassis lubrication points. Greases containing more than 3 percent should probably not be used in high-speed rolling bearings.

Molybdenum disulphide can also be incorporated in metallic or polymeric composites, which can be machined to produce bearings or other low-friction components. The coefficient of friction then tends to be higher, perhaps 0.1–0.3, depending on the composition, the load and the speed.

Molybdenum disulphide begins to oxidize at 350 °C in air, although it can still be used for short periods up to 450 °C. The oxidation produces molybdic oxide (MoO_3) which is itself a fair lubricant at higher temperatures but wears rapidly. In vacuum the problem of oxidation does not arise, and some reports have claimed satisfactory lubrication up to 1000 °C in high vacuum.

The maximum load-carrying capacity depends on the form in which the molybdenum disulphide is used, but it can be extremely high. In a four-ball machine (see Chapter 11) it is possible to extrude a hardened steel ball through a small orifice without seizure using a grease containing a high concentration of molybdenum disulphide.

Due to its poor thermal conductivity the maximum sliding speed is limited. The limiting PV also depends strongly on the form in which it is used, but may be of the order of 3.5 MN/ms (100 000 psi × ft/min).

The life in vacuum is determined by the film thickness and the wear rate. In air it has been shown that molybdenum disulphide oxidizes, even below 350 °C, presumably due to the effects of flash temperatures caused by friction. The coating will fail when it loses cohesion because of oxidation.

Apart from oxidation, it is stable to most chemicals, but is attacked by strong oxidizing acids and by alkalis.

On the whole, molybdenum disulphide is a very versatile and useful material where oils or greases cannot be used or do not have sufficient load-carrying capacity. Its main properties can be summed up as follows:

– low friction 0.03–0.1, depending on load;
– maximum PV probably about 3.5 MN/ms;
– excellent adhesion, especially when dry;
– usable at low temperatures and to 350 °C in air;

- excellent performance in vacuum;
- temperature limit in vacuum approaches 1000 °C;
- very high load-carrying capacity;
- black and, therefore, unacceptable for some processes.

8.5 Other inorganics

Many other inorganic compounds have useful properties for lubrication, but none are widely used. This is due to poor film-forming properties, or complicated application techniques, cost, or some other problem. Some of the compounds, and their important characteristics, are summarized in Table 8.4.

Table 8.4 Characteristics of some inorganic solid lubricants

Material	Characteristics
Calcium fluoride	Usable 400–900 °C. Low friction, low wear rate
Graphite fluoride	Usable up to 470 °C. Low friction, fairly low wear rate, slightly abrasive
Molybdenum trioxide	Usable 300–1000 °. Friction fairly low, adhesion fair
Boron nitride	Usable 300–950 °C (1700 °C if no oxygen present). Friction about 0.2–0.3
Lead monoxide	Usable 200–700 °C. Friction and wear rate low

8.6 PTFE and similar polymers

PTFE (polytetrafluoroethylene) is a polymer of tetrafluoroethylene, which in turn is ethylene gas (C_2H_4) in which all the hydrogen atoms have been replaced by fluorine.

$$C_2H_4 \rightarrow C_2F_4 \rightarrow$$

Ethylene

Polytetrafluoroethylene

It is often referred to as Teflon, but this is simply a trade name for PTFE manufactured by Dupont, just as Fluon is the trade name used by ICI.

It is a white solid with a slightly waxy appearance, hard to the touch but easily cut and deformed.

It is usually fabricated by sintering or hot compression moulding at above 300 °C. It is difficult to form large components accurately, but they can be finish machined without difficulty, provided that the cutting speed and depth of cut are kept low to avoid thermal expansion. Bonding PTFE to metals has always been difficult, but it is said that treatment with hot alkali gives a surface which can be bonded. Adhesives are commercially available which are claimed to be suitable. It should also be kept in mind that toxic and irritant vapours can be emitted from PTFE and other fluorinated polymers if the temperature rises during machining.

It is stable in use to almost 300 °C, but it changes its state at 325 °C and cannot be used above that temperature. It can also be used at very low temperatures, even down to −200 °C or lower in liquefied gases. There is, however, some evidence that the wear rate increases at very low temperatures. It is also very resistant to oxidation, and can be used for lubrication or sealing in oxygen systems.

The attractive forces between the long straight-chain molecules are low, so that PTFE has fairly poor mechanical strength. For the same reason the long molecules slide easily over each other; it is this which gives it its low friction properties.

In view of this stability and low atomic forces, it is not clear why PTFE forms good, smooth, strongly adhering films to metal surfaces; but it does – probably more readily than graphite.

PTFE is probably the most widely used of all polymers in sliding applications, because of its exceptionally low friction. In dry sliding against metals the friction of pure PTFE can be as low as 0.03 at high load and low speed; it probably rises to 0.1 or more at low load and high speed.

The maintenance of low friction and wear depends on the establishment of a smooth film on both surfaces. Such a film is very readily produced by transfer onto clean metal surfaces which slide against PTFE. The presence of liquids interferes with transfer film formation, and although quite low friction can still be obtained, the wear rate in the presence of liquids can be too high to be acceptable. However, in many large-scale civil engineering applications, such as bridge bearings, PTFE composites have been found to give excellent lubrication and wear life, even when immersed in water.

Even under ideal conditions the wear rate of pure, or unfilled, PTFE tends to be very high. The use of fillers or reinforcement can reduce the wear rate by a factor of 10 000. Useful reinforcing fillers include glass fibres, carbon fibres, and molybdenum disulphide. With all of them, however, the coefficient of friction increases, commonly up to 0.2. Unfilled PTFE also yields slowly under load, and will extrude out of a loaded bearing. The maximum PV is probably about 60 kN/ms. The use of fillers increases its modulus and yield stress so that such deformation is reduced or eliminated.

Two types of reinforced PTFE are very widely used. In one, Glacier DU, the PTFE is incorporated in the pores of a sintered bronze strip which has a steel backing strip. In the other, Ampep Fiberslip, the PTFE in the form of fibres is interwoven with glass fibre and the whole three-dimensional weave is impregnated with phenolic resin, which also fixes it to a steel backing strip. There is also a cheaper, lower temperature material called Fiberglide, similar to Fiberslip but with the glass fibre replaced by a synthetic textile fibre.

The maximum PV for all these three materials is probably about the same, at about 3.5 MN/ms (100 000 psi × ft/min). However, the woven types can carry higher loads, while the copper-reinforced DU can operate at higher speeds. This is due to the better thermal conductivity of the copper matrix.

There are many other different commercially available composites containing PTFE. Several are listed in Table 8.5.

Table 8.5 Some commercial PTFE composites

Type	Filler/reinforcement	% Filler	Friction	PV limit (kN/ms)
Rulon A			0.12–0.24	500–700
Fluorocomp 103	Glass fibre	15	0.14	350–500
Fluorocomp 174	Glass fibre molybdenum disulphide	15 5	0.14	400–600
Fiberslip	Glass fibre		0.10–0.20	3500
Glacier DU	Sintered bronze lead powder		0.10–0.20	3500

The white colour of PTFE is an advantage in certain industries, such as textiles, where the dark colour of graphite and molybdenum disulphide is unacceptable.

Several other polymers are similar to PTFE in certain respects. They all have higher friction, but give some compensation in lower wear rate, higher structural strength, or easier bonding to metals. None of them is as readily available or widely used as PTFE. The properties of PTFE and some of these other polymers are compared in Table 8.6. To sum up, the important characteristics of PTFE are:

– very low friction, from 0.03 to 0.1;
– high wear rate in unfilled state;

Table 8.6 Properties of some fluorinated polymers

Polymer	Friction	Max. temp	Characteristics
Unfilled PTFE	0.04–0.12	290 °C	Low wear resistance; hard to process
Filled PTFE	0.1–0.2	290 °C	High wear resistance; hard to process
PTFCE	0.12–0.3	200 °C	Fairly easy to process
Filled PVF$_2$		150 °C	Wear resistant; easy to process
Filled PVF$_2$/TFE	0.10–0.13	150 °C	Wear resistant; easy to process

- deforms slowly under load in unfilled state;
- reinforcing fillers reduce wear rate and deformation, but increase friction;
- usable from −200 °C to +300 °C;
- highly resistant to chemical attack;
- white colour useful in some applications;
- easily machined.

8.7 Nylons

Nylon is a general name given to a group of linear polymers, chemically known as polyamides. The five common types are nylon 6, nylon 6/6, nylon 6/10, nylon 11, and nylon 12. Their properties differ to some extent, but some general remarks can be made about nylons as a whole.

They are tough polymers, with high mechanical strength, which will creep, or permanently deform, under continuous load, but to a much lower degree than unfilled PTFE. Their mechanical properties can be very much improved by fillers or reinforcing fibres. They have good resistance to chemical attack, although not as good as PTFE or FEP. They are also susceptible to moisture absorption, which makes them swell and soften. The resulting reduction in strength is very considerable – by a factor of almost 3 to 1 for nylon 6/10 and over 5 to 1 for nylon 6.

Nylons form smooth transfer films on metal counterfaces, and when used unlubricated their friction is fairly low – from 0.1 to 0.2 at low speed, depending on load and temperature. Their wear rate is also quite low at low speed, being perhaps fifty times lower than for PTFE. At higher speed, however, above about 1 m/s, the friction and wear increase rapidly because of surface melting.

Bearing performance is markedly improved by the use of fillers such as PTFE and molybdenum disulphide, and reinforcement such as glass or carbon fibre. Bearings have to be designed with large clearances to cope with the effects of moisture swell and thermal expansion.

Nylons are especially suitable for use where liquids are present, and are widely used with water or oil lubrication, where the moving parts are small, lightly loaded, or fairly slow moving. In low viscosity liquids like water they have advantages over metal components in that there is little tendency to gall or seize when surfaces rub against each other. The liquid cools the surfaces so that they can operate satisfactorily at higher speeds than when dry.

Nylon is often used for small gear sets (e.g. car window winders and windscreen wiper motors) and bearing bushes and sliding bearings (e.g. door locks). This is: (a) because of ease and cheapness of manufacture, and (b) because the tolerances are large, so that no finish machining is needed. Such components are usually grease-lubricated so that technically this is not solid lubrication. However, they will often continue to operate adequately, even when grease is no longer present.

Some relevant properties of nylon are listed in Table 8.7.

Table 8.7   Some properties of different nylons

Type of nylon	Friction	Wear resistance	Max. temp.
Unfilled nylon 6	0.2–0.3	Fair	100 °C
Filled nylon 6	0.2–0.4	Good	150 °C
Unfilled nylon 6.6	0.1–0.28	Poor	130 °C
Filled nylon 6.6	0.2	Fair	200 °C
Unfilled nylon 6.10			110 °C
Filled nylon 6.10			200 °C

8.8   Acetals

Acetals are similar in their use to nylons, being more often used lubricated than unlubricated. In dry sliding the coefficient of friction is between 0.1 and 0.4 but can be reduced to 0.07 by using PTFE as a filler. The wear rate is also similar to that of nylons, and is improved by fillers and reinforcements. The temperature limit is only about 110 °C.

In lubricated applications acetals are outstanding, especially with regard to wear rate. This is so even when the speeds are low and loads high, and means that continual boundary conditions exist. But they also have the advantage that they are not damaged when lubrication fails, so they can be recommended for applications where the lubrication is sparse or intermittent.

Their main disadvantage is that they are difficult to attach to metal components. This makes them difficult to incorporate into a design. Where possible it is best to use commercially available bearings such as Glacier DX, in which the acetal is mechanically locked into a porous sintered bronze layer.

8.9   Polyetheretherketone (PEEK)

Polyetheretherketones are polymers with very good thermal and tribological properties, as well as other good mechanical properties and chemical inertness. In dry sliding the coefficient of friction of the unreinforced injection-moulded material varies from 0.15 at low speed and 200 °C, to 0.58 at high speed and 20 °C. The bearing grade material is reinforced with carbon fibre and solid lubricant such as graphite or PTFE. It has a coefficient of friction in dry sliding between 0.11 and 0.17 over a wide range of temperatures and sliding speeds.

The wear rate of the same reinforced grade is less than that for similarly reinforced nylon but higher than for reinforced acetal and polyimide. The limiting PV in dry sliding is claimed to be over 10 MN/ms. The limiting temperature for continuous use is 260 °C and, for brief periods, 300 °C.

PEEK is easily processed by injection moulding and can be attached to metal components. It is very resistant to attack by most materials, so there seems to be no reason why it cannot be used with conventional lubricants, but specific information is lacking.

8.10   Other polymers

No other polymers are widely used as dry lubricants or dry bearing materials. For slow, lightly loaded sliding contacts many are useful because they give fairly steady friction, low wear rate, and have a much lower tendency to seizure than metal against metal. In general, however, they have no advantages over acetal, nylon, or reinforced PTFE except at very high temperatures, where polyimide, polyphenylene sulphide, and polysulphones can be useful.

Several polymers, such as phenolics, epoxies and polyimides, are used as binders for molybdenum disulphide or other dry lubricants in coatings. In addition, for various reasons, several different polymers are used for lubricated bearings, with either oil, grease or water. Their dry rubbing properties may become important if the bearings are starved of lubricant.

The following paragraphs summarize a few relevant points about some of the many other polymers. It should be remembered that almost any polymer can be strengthened by reinforcement and its friction reduced to 0.1 or less by incorporating some PTFE in it.

Polyimide
Usable continuously to 350 °C under high load, coefficient of friction 0.1 to 0.2 and steady. They are excellent bearing materials, but they are difficult to process and are generally only used where their high-

temperature capabilities are important. Have been used as binders for molybdenum disulphide coatings.

Polyethylene
Temperature limit about 75 °C, and attacked by solvents. High-density form (HDPE) is used to some extent for bushes in water, but they are damaged if they run dry. Ultra-high molecular weight form (UHMWPE) is used for 'bearings' in artificial hip joints.

Polysulphone
Temperature limits vary from 150 °C to 260 °C. Coefficient of friction 0.14 to 0.22. Could be used as binders for coatings, and could replace nylon for many purposes. At present not widely used.

Polyphenylene sulphide
Temperature limit over 250 °C. Friction rather variable, 0.2–0.5. Performs well with water lubrication but not when dry.

Phenolic
Used very widely with fabric reinforcement for lubricated bearings, with either oil, grease or, especially, water. Also widely used as a binder for PTFE or molybdenum disulphide in coatings and for PTFE in composites. Mechanical properties very good. Without lubricating fillers not very useful in dry sliding, and the friction varies from 0.16 to 0.5 or more.

Epoxy
Epoxy resins are used as binders for molybdenum disulphide coatings and for composites. Also used in the same way as phenolic resin for water or grease-lubricated bearings, but not as widely. Better than phenolics in dry sliding, with friction 0.2 and high-wear resistance. Would not normally be deliberately used for dry rubbing applications.

8.11 Metals as solid lubricants

It was explained earlier that the friction of a material depends, to some extent, on the force needed to shear it. Soft metals like lead can be sheared easily, so that in certain circumstances they can also be used as solid lubricants. Being soft, they will not support heavy loads without deforming, so they are used as thin coatings on stronger metal surfaces.

The coefficient of friction is not low, being usually about 0.3, but the films are stable. They can be used to quite high temperatures, and have the additional advantage that they are good conductors of heat and electricity. Table 8.8 gives the melting points of some of the metals which are of most use as lubricants.

At low speed and load the metals can be used at temperatures

Table 8.8   Properties of some lubricating metals

Metal	Melting point	Hardness (Moh scale)
Gallium	30 °C	1
Indium	155 °C	1
Tin	232 °C	1.8
Thallium	300 °C	1.2
Lead	327 °C	1.5
Barium	704 °C	
Silver	961 °C	2.5
Gold	1063 °C	2.5

approaching their melting points, but at higher speed or load frictional heating increases the actual surface temperature and local melting may take place.

Apart from silver and gold, metal films are mainly used as lubricants in vacuum, and it is presumed that oxidation in air interferes with smooth sliding. In tinplate manufacture and forming, the tin on the surface provides useful lubrication, but this is only needed for relatively short periods.

Films are usually about 0.25–1.0 μm (10–40 μin) thick, and must be very smooth and uniform. The friction is found to increase with either thicker or thinner films.

Silver and barium films have been used successfully on lightly loaded ball bearings in high vacuum X-ray tubes. Silver and gold have been tested successfully for use in high vacuum in spacecraft applications. Perhaps the most successful applications so far have used thin films of lead on ball bearings for high vacuum, such as in spacecraft.

Lead and copper powders are also used in anti-seize compounds for high-temperature use, where their softening at high temperature further reduces the friction.

8.12 Composites

Any solid materials which contain two or more different solid phases mixed together are known as composites. Common examples are 'fibre-glass', which contains glass fibres dispersed usually in an epoxy or phenolic resin, and most rubbers, which contain carbon black or other fillers as well as the rubber polymer.

For bearing use most polymers are strengthened by glass or carbon fibre, and many of them have PTFE or molybdenum disulphide added to reduce friction. Table 8.9 lists some of the materials which can be used in composites. Since the number, types, and concentration of different

Table 8.9   Materials used in lubricating composites

Matrix	Lubricant	Reinforcement powders	Filler fibres	Additives
PTFE	PTFE	Graphite	Glass	Antimony trioxide
Nylon	Molybdenum disulphide	Carbon	Carbon	Lead phosphite
Acetal	Graphite	Molybdenum disulphide	Steel	Sodium nitrite
	Calcium fluoride			
Polyimide	Lead	Lead	Copper	Alumina
Bronze	PTFCE	Copper	Terylene	Butyl phthalate
Cast iron	Niobium diselenide	Silica	Nomex	
Nickel	Lead monoxide	Alumina	Rayon	
Silver	Boric oxide	Mica		

substances used can all be varied, it is obvious that the variety of different possible composites is enormous. A few examples will be enough to indicate the way in which composites are formulated.

Glacier DU

Briefly described earlier, Glacier DU consists of a porous sintered bronze strip with a steel backing. The pores of the sintered bronze are filled with PTFE in which lead powder is dispersed, and a thin layer of lead-filled PTFE coats the surface. The steel backing strip gives structural strength and maintains the bearing shape. The sintered bronze strip supports and reinforces the PTFE and gives high thermal conductivity. The PTFE provides low friction, and the lead powder reduces the wear rate of the PTFE and appears to improve transfer film formation.

Deva metal

This is, in fact, the general name for a series of rather simple composites, each containing only natural graphite (4–14 percent by weight) and a metallic matrix (either cast iron, nickel, leaded bronze, or tin bronze). The complexity arises in the manufacture. The graphite and metal powder are thoroughly mixed. It seems probable that, in mixing, the graphite forms an adherent film on the surfaces of the metal powder. The powder is then compressed in a mould and sintered. This gives a strong metal matrix whose pores are filled with graphite, and which releases graphite as a dry lubricant during use. The metal matrix again gives high thermal conductivity.

Duroid 5813
Typical of a large number of commercial composites which consist of PTFE with glass fibre reinforcement (15–25 percent) and molybdenum disulphide (3–5 percent). The glass fibre gives high structural strength, while the molybdenum disulphide reduces the wear rate and improves transfer film formation.

8.13   Selection of solid lubricants

The selection of solid lubricants falls basically into two parts. The first involves the decision as to whether solid lubrication should be used instead of oil, grease, or gas lubrication. The second involves deciding which is the most suitable of the available solid lubricants.

The first decision depends on the advantages and disadvantages of solid lubricants as compared with oils and greases, which were summarized in Section 8.2. It follows from the advantages listed that solid lubricants may be the preferred choice when one or more of the following factors applies:

(a)   temperature too high for oil or grease;
(b)   temperature too low for oil or grease;
(c)   high vacuum environment;
(d)   important to avoid contamination, e.g. textiles or electrical contacts;
(e)   reactive chemical or radioactive environment;
(f)   equipment needs to operate quickly and reliably after long storage, e.g., safety devices or missiles;
(g)   loads are high, with little or no sliding speed;
(h)   vibrating contacts.

Even where these factors apply, the decision may not be an obvious one, because a system can often be redesigned to allow oil or grease to be used. For example, additional heating or cooling may allow an oil or grease to be used; sealing may protect the system from the high vacuum or reactive chemicals; shielding may reduce the radiation and so on. As in all equipment design the choice then becomes a matter of balancing the advantages and disadvantages of the different possible solutions.

In the end the decision to use a solid lubricant will often be based on the simplicity of design possible with it, as compared with the complexity needed to cope with oil or grease in adverse conditions.

The second part of the selection process depends on the properties of the individual solid lubricants. An attempt has previously been made to summarize those properties which are likely to influence the selection. It must be remembered, however, that compromises can be obtained by combining two or more of them in a composite.

8.14   Designing for solid lubricants

Solid lubricants wear away in use. This limits their service life, unless some system of replenishment can be arranged. In addition wear causes an increase in bearing clearances, play, backlash, etc. The problem of replenishment will be considered later, but in most applications replenishment will not be possible, and all the lubricant required must be incorporated at the beginning.

Being solids, they have shape or form, and an important part of designing for solid lubricants is to decide what shape or form is required, how it is to be produced, and how it is to be built into the system.

(a)   Wear life

Two properties of solid lubricants must be known in order to plan their performance and wear life.

The first is the combination of speed and load at which they can be used. The PV factor was explained in Section 8.1, together with the need to quote the maximum permissible pressure and the maximum permissible sliding speed. These last two properties are not often quoted, and reliable figures are hard to obtain. Some figures for limiting PV, pressure, and sliding speed are listed in Table 8.10, but these may not be very accurate.

The second important property is the wear rate. It has been found that with certain solid lubricants and many other materials which wear smoothly, the volume of material removed by wear depends only on the

Table 8.10   Approximate speed and load limits for solid lubricants

Solid lubricant	Limiting PV $(KNm^{-2} \times ms^{-1})$	Maximum speed (ms^{-1})	Maximum pressure (MNm^{-2})
Graphite	700	0.5	70
Molybdenum disulphide	3500	0.2	2000
Unfilled PTFE	60	0.01	10
Filled PTFE	500	0.02	100
Reinforced PTFE	3500	0.1	400

contact load (not pressure) and on the total sliding distance. This rule applies as long as the limiting PV, pressure, and speed are not exceeded.

If the volume removed is proportional to the load and the sliding distance, we can write

$$\text{Volume (mm}^3) = k \times \text{load (newtons)} \times \text{distance (metres)}$$

where k is a constant called the specific wear rate. This equation can be written

$$k = \frac{\text{volume}}{\text{load} \times \text{distance}}$$

where k is written in mm^3/Nm. It has become increasingly common in recent years to quote the specific wear rates for different solid lubricants.

If the total load on a bearing, the contact area and the specific wear rate are known, the depth of wear after a certain sliding distance can be calculated. For example, suppose a bonded molybdenum disulphide film for use in vacuum is 50 μm thick with a specific wear rate of 10^{-5} mm^3/Nm, a contact load of 20 N, contact area 6 $\times$ 5 mm and sliding speed 0.1 m/s, then

$$\text{Volume removed} = 10^{-5} \times 20 \times \text{sliding distance}$$

The film will be completely removed when the contact area 5 $\times$ 6 mm is worn down 50 μm, so that

$$\text{Volume removed} = 15 \times 6 \times 0.05 \text{ mm}^3$$

$$= 1.5 \text{ mm}^3$$

The sliding distance will then be given by

$$\text{Sliding distance} = \frac{1.5}{20 \times 10^{-5}} \text{ metres}$$

$$= 0.75 \times 10^3 = 7500 \text{ metres}$$

At a sliding speed of 0.1 m/s this distance will be covered in 75 000 seconds or 20 hours and 50 minutes.

A safety factor should be applied to the design to allow for the fact that the wear would not be completely uniform and that the specific wear rate is only approximate. It might, however, be reasonable with these figures to expect a wear life of 10 hours or more.

For a service life of 150 hours, the simplest approach might be to increase the contact area to 20 mm square, or to use a stronger film with a specific wear rate of 10^{-6} mm^3/Nm.

This type of calculation is valid for molybdenum disulphide in vacuum or for many other types of solid lubricant in air. As pointed out earlier, the life of a molybdenum disulphide in air may be limited by oxidation.

In the example quoted, the bearing life was limited by the thickness of the bonded film. With a thicker film the life might not be much longer, because thicker films are often weaker, and the maximum permissible thickness for bonded or burnished films is usually about 50 μm.

If the lubricant is incorporated in a polymer, thicker films or even bulk solids can be used. The wear limit is then likely to depend on the amount of increased play or slackness which can be accepted.

(b) Applying the solid lubricant

If the solid lubricant is to be used in the form of a bonded film, then the suppliers' instructions about pre-treatment, application, and film thickness should be followed. Bonded films are similar to paint films, and can be applied in similar ways, but the thickness required for effective lubrication is much thinner than for a good paint film.

Burnished films can be applied up to 10 μm thick. Sputtered films are generally thinner and stronger, but can only be applied by specialists.

Where the solid lubricant is in the form of bulk solid, such as nylon or reinforced PTFE, the problem is to machine the component to the required shape and incorporate it into the assembled system.

Machining of most of the lubricating polymers is straightforward, provided that the material is kept cool to avoid thermal distortion. It is, however, subject to the problems of toxic and irritant vapours which were mentioned earlier. Assembly into the equipment is sometimes more difficult because of low structural strength and differential thermal expansion. This is why many commercial lubricating composites are supplied with a steel backing strip.

So far it has been assumed that solid lubricants cannot be replenished in service. Generally this is true, but a few techniques have been used.

Molybdenum disulphide powder has sometimes been supplied to ball bearings and gears by means of an air jet, but the technique has not been widely used.

Both molybdenum disulphide and graphite can be supplied with grease or oil to a bearing, but in these cases they are acting as additives rather than simply as solid lubricants. It is possible to supply them in a volatile solvent, which then evaporates, but it is not easy to get a uniform spread. Similarly, a dispersion of graphite in a polyglycol can be fed to a very hot bearing (between 200 °C and 500 °C). The polyglycol decomposes and volatilizes away, leaving the graphite as a solid lubricant.

The most useful replenishment technique is the transfer technique used mainly with PTFE or molybdenum disulphide. With this technique a block of PTFE or molybdenum disulphide composite in a suitable form is mounted in a position where it is in sliding contact, directly or indirectly, with the parts which require lubrication. These parts thus become coated with the solid lubricant, which eventually coats all the sliding surfaces. With suitable design the transfer technique can provide replenishment of the solid lubricant for a long period.

8.15 Some applications of solid lubricants

In terms of numbers, the most important applications of solid lubricants are domestic rather than engineering, with graphite in pencils and PTFE in non-stick frying pans and other utensils. PTFE has also been used for artificial ski-slopes and for coating skis. Nylon is used for bearings and gears in washing machines, windscreen wipers, car window winders, and many other consumer products, although it is common to use a grease or oil with it, at least initially.

In engineering, solid lubricants are used in almost every type of component, including plain and rolling bearings, slider bearings, gears, seals, electrical brushes, brakes, and clutches.

(a) *Plain bearings*

PTFE is probably the most widely used solid lubricant for plain bearings. Many of these are simple cylindrical bushes. Unfilled PTFE may be used for lightly loaded bushes and reinforced PTFE for more heavily loaded ones. The more complex PTFE composites and woven forms are used in highly critical situations, such as support bearings of bridges, and control surface and variable geometry wing bearings in high-performance aircraft. PTFE is also used to lubricate the tracks when bridges are slid bodily into position to minimize interruption of road or rail traffic.

Molybdenum disulphide tends to be used on sliding surfaces of all sorts of shapes, because it can be easily coated onto the basic metal surfaces. It is also used as a filler in some PTFE bearings.

(b) *Rolling bearings*

Many different techniques have been tried for lubricating ball and roller-bearings with solid lubricants. The most common now is to use a reinforced PTFE cage or separator, possibly with molybdenum disulphide as a filler. The cage is rubbed by the balls or rollers, which become coated with a transfer film or PTFE, and in turn transfer it to the races.

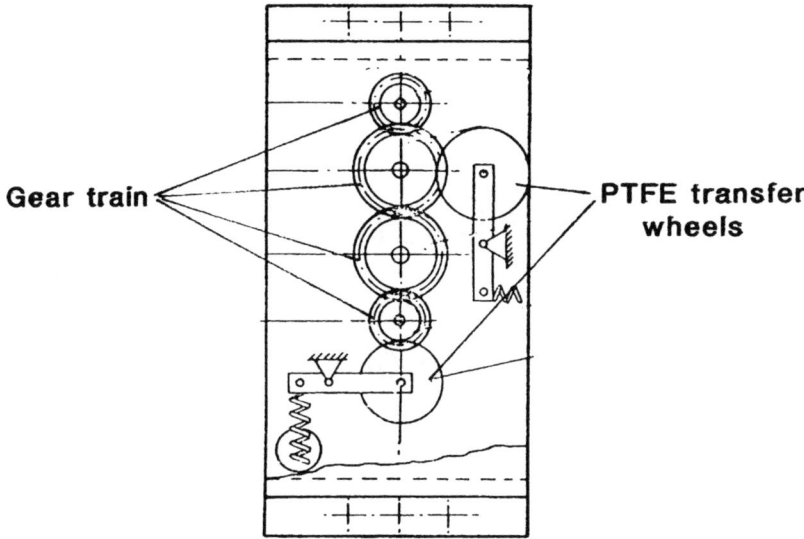

Figure 8.4   Transfer lubrication of gear train

(c)   *Gears*

The best technique for solid lubrication of gears is also to transfer the lubricant to the gear teeth. Figure 8.4 shows a Russian design for transferring PTFE or molybdenum disulphide to gear teeth from an idler gear made of a suitable composite.

(d)   *Electrical brushes*

The traditional solid lubricant for electrical brushes is carbon-graphite. The use of molybdenum disulphide as an additive to carbon-graphite has been mentioned above, but for space use one of the best materials is a composite containing 82.5% silver, 2.5% copper, and 15% molybdenum disulphide.

(e)   *Brakes and clutches*

The friction of brakes and clutches is not necessarily particularly high. A typical figure for the coefficient of friction is between 0.3 and 0.4. It is important, however, for the wear rate to be low and for the friction to be constant and not badly affected by heat. In order to achieve this, brake composites contain small amounts of graphite or molybdenum disulphide to control the friction.

Appendix

Theory of friction of solid lubricants

The frictional force, F, between two solids is roughly equal to the critical shear stress, S, of the softer of them, multiplied by the real area of contact, A, between the two. (This assumes that the friction is largely adhesional, see Chapter 1.) In lamellar solid lubricants the sliding ideally takes place between adjacent lamellae (flat crystals or sheets of straight polymer chains) oriented parallel to the sliding direction.

If there is no tangential force or motion, the real area of contact, A_n, with a ductile material is equal to the normal force, W, loading the two surfaces together, divided by the plastic yield pressure p. If a tangential force is applied to one of the components, the force which has to be supported is the resultant of the normal force and the tangential force, and the real area of contact increases. This phenomenon is known as junction growth.

A practical limit A_1 to the real area of contact with junction growth is often between $3A_n$ and $6A_n$.

The yield pressure p, or hardness for many materials is typically about five times the critical shear stress, so the coefficient of friction will be

$$\mu = \frac{F}{W} = \frac{A_1 S}{A_n P}$$

and since $p = 5S$ and A_1 is (3–6) A_n'

$$\mu = 0.6 - 1.2.$$

Adhesion and junction growth can take place between the solid lubricant and a metal substrate, so that the friction will be high.

Between two layers of dry lubricant, however, the yield pressure, p, is high compared with the very low critical shear stress, S, and junction growth probably does not occur. If p is as much as $10S$, and $A_1 = A_n$ we will have

$\mu = 0.1$

In practice p must often be greater than $10S$, because the actual coefficient of friction is even lower under ideal conditions.

Chapter 9

Gas Bearings

9.1 Principles of gas bearings

A gas can be used as a lubricant in much the same way as an oil is used in a hydrodynamic bearing. The main differences arise from the fact that a gas has a much lower viscosity than a liquid.

The theory of hydrodynamic lubrication was explained briefly in Chapter 1. The situation can be summarized by saying that, for a given surface roughness, the load which can be carried increases with the lubricant viscosity and the speed. Due to the very low viscosity of a gas compared with an oil, 'gas dynamic' bearings generally differ from hydrodynamic bearings in the following ways.

(a) the operating speeds are higher,
(b) the loads are lower,
(c) the surface finishes are better (i.e., smoother),
(d) the clearances are smaller, both to optimize load-carrying and to reduce gas flow,
(e) as a result the bearing tolerances must be very low.

The load carried by a gas bearing can be increased if the gas is pumped in under high pressure, so that the externally-supplied pressure carries the load. Such bearings are known as 'externally-pressurised' bearings. Since they rely on the external pressurization rather than speed to provide the load-carrying capacity, they may differ from gas-dynamic bearings in that:

(a) they can operate more slowly, or even stationary,
(b) they can have poorer, (i.e., rougher) surface finish, although usually they do not,
(c) they can carry higher load.

However, for efficient operation it is still necessary to reduce the volume of gas required to a minimum. As a result, in practice externally pressurized gas bearings also have small clearances, tight tolerances, and high quality surface finishes. These factors are less critical than for gas-dynamic bearings.

The basic design of the bearing is not much more difficult than for oil lubrication, but it is complicated by the requirements for smooth surfaces and tight tolerances. It follows that gas bearings represent a very high level of precision engineering, and their design and manufacture tends to be done by specialists.

Gas-bearing technology is in fact largely a matter of design and manufacture, which are outside the scope of this book. The remainder of this short chapter will be concerned with the factors involved in deciding when and how to use gas bearings.

Before leaving the subject of gas bearing types, it may be helpful to mention the various names which are used for them. There is no clear agreement about the names except for the cumbersome terms 'self-pressurizing' and 'externally pressurized' gas bearings. The following alternative names are used:

(i)   self-pressurizing gas bearings
      aerodynamic bearings
      hydrodynamic gas bearings
      gas dynamic bearings
(ii) externally pressurized gas bearings
      aerostatic bearings
      hydrostatic gas bearings
      gas-static bearings

9.2 Properties of the gas

Any gas can be used as the lubricant in a gas bearing, provided that it is clean enough and does not react with the other materials in the system.

Solid contaminants can cause two problems. Large particles may be caught between the bearing surfaces, causing damage to the surfaces and disturbing the stability of motion. Externally pressurized bearings may have clearances of the order of 25 μm (0.001"), similar to those in oil-lubricated bearings, so that filtration to a nominal size of 10 μm may be adequate.

The second problem with solid contaminants is that they can cause erosion of surfaces, especially where the flow rates are high, such as at orifices. Erosion is related to the total concentration of particles and their size and abrasiveness. Erosion can be reduced by avoiding sharp changes in the direction of gas flow.

With self-pressurizing gas bearings the clearances may be very much smaller; consequently, greater cleanliness is needed. On the other hand, the volume flow rates are generally low, and the air will often recirculate within the housing so that a high level of sealing can be used. It is also possible to design the inlet with a restriction smaller than the bearing clearance, to exclude damaging particles.

Liquid contaminants can cause problems if they settle out on the bearing surfaces or on high-speed components, causing either sticking or imbalance. The same can apply to vapours which might condense out in the bearings, so that for example if dry steam is used as the operating gas it must be kept well above its condensation temperature.

The use of gas lubrication removes one of the major restrictions on liquid lubrication, namely the restricted temperature range of use. All substances which are liquid at normal temperatures will be solid at $-100\,°C$ and gaseous (or decompose) at $600\,°C$. Gases can exist at any temperature provided the pressure is low. Even at pressures high enough to be practical for gas bearings some of them exist as gases at any temperature above $-200\,°C$. A substance will exist as a gas whatever the pressure if the temperature is above its critical point. Table 9.1 lists the critical points of a few common gases.

The viscosity of a gas increases with increase in temperature, as shown in Fig. 9.1. This effect is not enough to simplify the design problem at high temperatures. For example, even at $1000\,°C$ the viscosity of nitrogen is still less than 0.1 cP, or one tenth of the viscosity of water at normal temperature. It does, however, increase the difficulty of designing self-pressurizing gas bearings for very low temperatures. There have been reports of the use of hydrodynamic gas bearings in cryogenic systems. Generally, however, externally pressurized gas bearings are likely to be preferred.

The choice of gas is not often a problem. It will usually be the one

Table 9.1 Triple point and critical temperatures

Liquid	Critical temperature °C	Critical pressure (bars)	Triple point °C
Carbon dioxide	+31	73	−57
Fluorine	−129	52.1	−218
Helium	−268	2.26	
Hydrogen	−240	12.8	−259
Nitrogen	−147	33.5	−210
Oxygen	−119	49.9	−219
Water	+374	217.7	+0.007

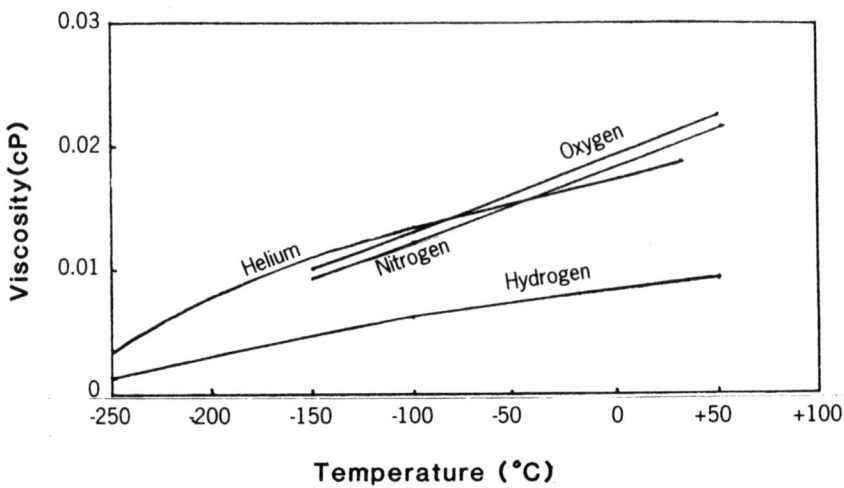

Figure 9.1   Increase in viscosity of gas with temperature
(Note: Dynamic viscosity in centipoise is shown in this figure because the kinematic viscosity in centistokes is almost meaningless for a gas.)

most readily available, and in fact the vast majority of gas bearings are lubricated with air. If some other high-pressure gas is being used in a system, then it may be very economical to use it as a lubricant. Many gases are used in large quantities nowadays, including nitrogen, oxygen, hydrogen, ammonia, methane, and propane. There is a great deal of scope for using them for lubrication of components used in handling them.

Apart from air, two gases sometimes chosen and supplied specifically for gas bearing use are nitrogen and helium. Nitrogen is readily available and relatively cheap, either compressed or in liquid form. It is chemically very inert, and is therefore safer to use than other readily available gases like oxygen and hydrogen, or even compressed air. It is sometimes used for externally pressurized gas bearings, especially in experimental equipment.

Helium is neither readily available nor cheap, but it is even more inert than nitrogen. It is sometimes used for self-pressurizing gas bearings in small sealed systems such as inertial gyroscopes.

It was said earlier that any clean gas can be used, provided it does not react with system materials. There are, of course, many gases which will attack common metals, and they would not normally be chosen for gas bearings. Where they are being manufactured or processed they may still be a good choice for lubrication because this avoids the problem of separating them from some other lubricant. It will then be necessary to choose inert bearing materials such as glasses or ceramics to avoid attack by the reactive gas.

9.3 Advantages and disadvantages of gas bearings

The main advantage of gas bearings has already been mentioned, namely the very wide temperature range over which they can be used, from perhaps −200 °C to 2000 °C. In practice it would be very difficult to construct a single bearing to cover a range of even 500 °C because of the problems of thermal expansion and maintaining clearances. This would not, however, be impossible, and designs such as conical self-pressurizing bearings might well cover much wider ranges.

Very high and very low temperatures can also be achieved with solid lubrication. The choice between gas and solid will generally be an easy one to make, since solid lubricants are suitable for low speeds and high loads, while gas lubricants are suitable for high speeds and low loads. In addition the friction of solid lubricants will usually be between 0.03 and 0.1, while that of gas bearings may be too low to detect.

It is theoretically possible to use a solid lubricant to manufacture a gas bearing, so that as the load increases and the gas film ceases to separate the surfaces, the solid lubricant can take over. The practical difficulty with this technique is that solid lubricants wear away with rubbing and it is difficult to maintain the very precise shape needed for a gas bearing. A compromise sometimes used is to coat the surfaces of a gas bearing with a very thin sputtered or burnished film of a solid lubricant. This will reduce friction and damage with the occasional 'touch-down' of the bearing, but the hard substrate will ensure that the shape is maintained.

Another advantage of gas bearings is the high bearing stiffness or rigidity which can be obtained. In most cases this is due to the very small clearances, or in other words to the very thin lubricant film, which ensures that the moving parts maintain very precise location. A secondary factor in very high-speed rotating gas bearings is their gyroscopic rigidity. Where the bearings are kept at the same orientation in space, as in inertial gyroscopes, this also assists the bearing stiffness. Where they are moved about, as in a dentist's drill, the opposite is true.

The advantages and disadvantages of gas bearings can be summarized as follows:

Advantages

(a) Wide temperature range
(b) High bearing stiffness
(c) Usable at very high speeds
(d) No requirement for special lubricant supply for self-pressurizing gas bearings

(e) Clean in use
(f) Can operate in completely sealed environment
(g) Exceptionally low friction

Disadvantages

(a) Generally low load-carrying capacity
(b) Detailed design complicated
(c) Precise control of shape and surface finish required
(d) Need very clean gas supply
(e) Bearing surfaces easily damaged by touch-down in stop-start operation or overloading

9.4 Examples of gas bearing use

(a) *Dentists' drills*

High-speed drills used by dentists have often had externally pressurized air bearings, and been driven by an air turbine. The arrangement is shown in Fig. 9.2, and includes a journal bearing and a thrust bearing. The journal bearing is an ideal application for a gas bearing, as the speed is very high, of the order of 100 000 rpm, and the loads are relatively low. The radial loads consist of minute out-of-balance loads together with the radial component of the drilling loads. The smaller

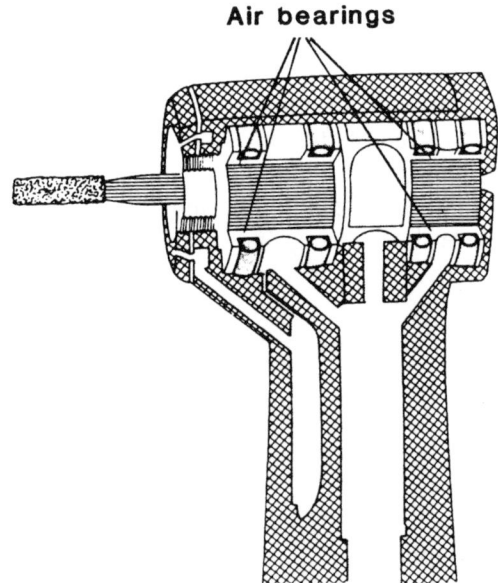

Figure 9.2   Externally pressurized air bearings in a dental drill

thrust bearing supports the axial component of the drilling loads, and is the critical part of the design.

The only supply required is the compressed air supply. This is by a rubber hose, so that the control of the drill is very flexible. The high bearing stiffness gives very precise control of the drill.

(b) *Precision grinding heads*

The high precision and stiffness of a gas bearing are the main reasons for their use in precision grinding heads. There is an externally pressurized journal bearing with a thrust collar to maintain axial position. The speed of rotation is relatively low, but the load is quite high. This makes it necessary to use an externally pressurized bearing rather than a self-pressurizing type. Apart from the higher load-carrying capacity of an externally pressurized bearing, this type of bearing allows the rotor to lift off before rotation starts, and thus avoid damage.

(c) *Inertial navigation gyroscopes*

The gyroscopes used in inertial navigation systems for aircraft and ships are required to operate for lengthy periods without any precession, or tilting of their axes. This is achieved by means of high gyroscopic rigidity, which in turn requires high speeds of rotation, such as 24 000 rpm. The gyroscopes are therefore mounted on small conical self-pressurizing gas bearings, as shown in Fig. 9.3.

Since they are self-pressurizing, the air film decreases in thickness as the bearing slows down after use, and the surfaces touch while they are still rotating. Similarly, they must start rotating again before their

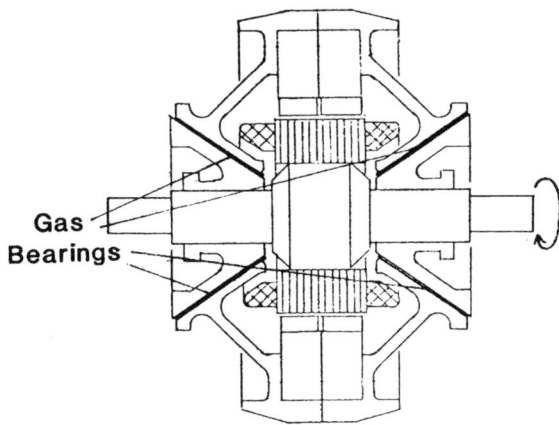

Figure 9.3 Gas bearings in an inertial gyroscope

surfaces are separated (before 'lift-off'). This makes it important to ensure that their wear resistance is high, so that they maintain their precise shape and surface finish. It is also important that the sliding friction is low, and remains low, so that they restart easily.

Experimental work has been carried out on the use of boundary lubricants, coated onto the bearing surfaces to reduce friction and wear. These are not yet widely used in service.

To avoid oxidation and corrosion, the gyroscopes are contained in sealed housings filled with dry helium. The bearings are therefore helium gas bearings and not air bearings.

The start and stop problem applies particularly to aircraft inertial navigation systems, since a typical flight will be between three and twelve hours. In a ship, the system may be operating continuously for several days or even weeks, and the problems associated with frequent starting and stopping will not apply.

Chapter 10

Sealing of Lubricants

10.1 Principle of sealing

Seals are used to prevent a fluid (i.e., a liquid or gas) from escaping from one place to another. The technology of sealing is very broad and often complex, and is covered very well in publications such as *The seal users handbook* (Austin *et al.*, 1974, BHRA Fluid Engineering). This chapter will only deal with those aspects which are relevant to the sealing of lubricants.

Often the same seal will succeed in confining the lubricant and in excluding other fluids, but this is not always the case. Many seals are effective in only one direction. To separate a lubricant from another fluid it may be necessary to use two or more seals 'back-to-back', each preventing movement of one of the fluids.

In general it is easy to prevent a fluid from escaping, provided that no moving parts are involved. Containers can be made of all sorts of materials, from paper and rubber, to glass and steel. Any openings can be effectively closed by lids or stoppers of the same wide range of materials. There are fluids which are hard to contain because they are strong solvents or corrosive, but some suitable materials for a container can always be found.

The problem becomes much more difficult when a moving shaft has to pass through the seal. This is the usual situation where lubricants are involved. The seal has to cope with movement between its own surface and that of the shaft. These surfaces are bearing surfaces, having the same need to control friction and wear as other bearings, but with the added need to ensure that the gap between them does not allow the sealed fluid to escape.

Many different techniques are used to cope with this problem of sealing moving components. They can be conveniently described in four groups: static; semi-static; rotating; and reciprocating. The last are known as dynamic seals.

A fifth technique is to avoid the need to separate a lubricant from another process fluid by using the process fluid as the lubricant, thus eliminating the sealing problem entirely. Probably the most widespread use of this fifth approach is in the so-called 'glandless pumps' used mainly in the chemical industry.

10.2   Static seals

One method for sealing a moving shaft efficiently is by completely enclosing the shaft in a stationary container, and driving it by means of a rotating magnetic or electromagnetic force.

The simplest method for achieving this is shown in Fig. 10.1. It involves a rotating magnet outside the container and a corresponding magnet attached to the shaft inside the container.

The internal magnet will orient itself so that the opposite poles of the two magnets are in proximity. As the outer magnet is rotated, the inner one also rotates to keep the opposite poles close together.

Any friction in the supports of the inner shaft will make it lag behind the outer one. The greater the lag, the lower is the magnetic force acting on the slave magnet. A situation can easily be reached with a particular

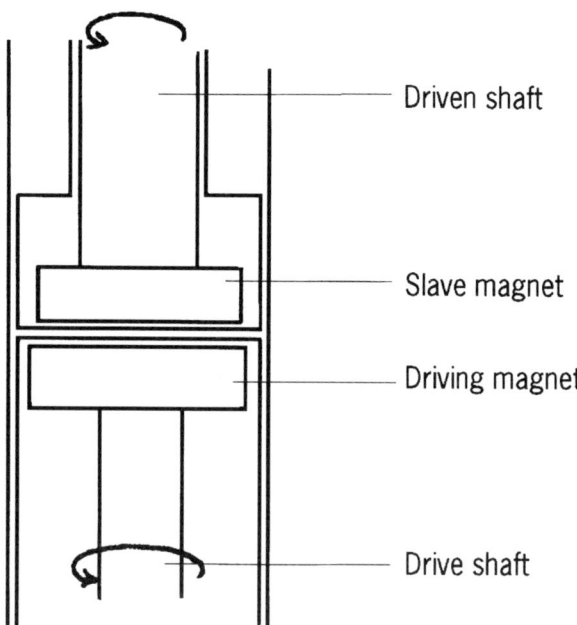

Figure 10.1   Magnetic drive through static seal

combination of drive magnet speed and inner shaft friction where the inner shaft fails to rotate.

This simple system has rarely, if ever, been used for precision machinery. It has, however, been used for magnetic stirrers, children's toys, and advertising demonstrations.

A more sophisticated system is to mount the field coil of an electric motor outside the container and the armature on the rotary shaft inside it, as shown in Fig. 10.2. This provides a much more precise and controllable machine. It has been used in spacecraft to protect the lubricant in the shaft support bearings from the effects of space vacuum. It has also been used in other high-technology applications, but never very widely.

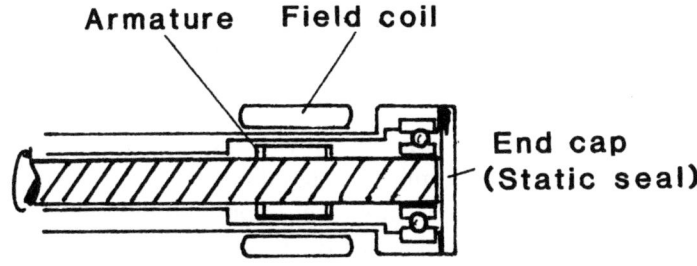

Figure 10.2   Split electric motor drive with static sealing

10.3   Semi-static seals

In some lubricated systems the movement which takes place is oscillatory and of low amplitude compared with the size of the components involved. In such a case, very effective sealing can be achieved by means of a flexible sleeve or shroud which is tightly fitted to the moving part or parts. There is then no movement between the surface of the seal and the surface of the component, and all the movement takes place in the material of the seal.

One of the simplest versions is used on multi-speed gearboxes such as those on vehicles or machine tools. Where the gear-change lever enters the gearbox it is common practice to fit a domed or conical rubber sleeve which is attached tightly to the lever and to the outside of the box. Two versions of this device are shown in Fig. 10.3.

A sleeve of this type can cope with a wide change in angle of the gear-lever. It may also be used to accommodate an increase or decrease in the volume of air and oil contained in the gearbox due to changes in temperature. In this way a gearbox can be completely sealed, without an air vent or breather having to be fitted.

The actual sleeve shape chosen will depend on the angle through

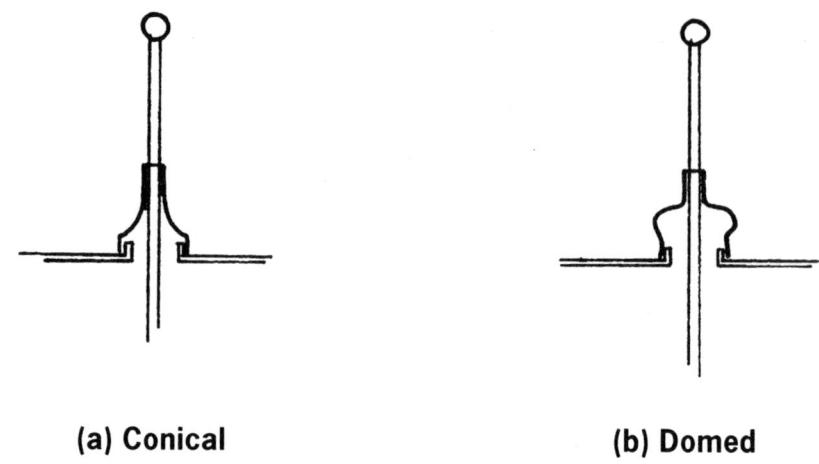

(a) Conical   **(b) Domed**

Figure 10.3   Rubber sleeve as static seal for gearbox

which the lever must be able to move, and on the extent to which
expansion and contraction must be handled. The form shown in Fig.
10.3 (a) permits large angular movements but little expansion, while the
one in Fig. 10.3 (b) allows a great deal of expansion.

Where the relative movement is entirely – or almost entirely – axial,
the best flexible seal may be a bellows. A typical example of such an
application may be at a splined coupling.

It is often necessary for a rotary drive to be applied through a
universal coupling, and the coupling may have to permit wide changes
in angle. A common example is in the drive to a rolling mill, especially
the early stages of a hot mill. A flexible sleeve called a 'muff' will
sometimes be fitted to the input and output shafts to surround the
coupling, as shown in Fig. 10.4. Such a seal will ensure that the
lubricant stays where it is needed, and will reduce the fire risk by

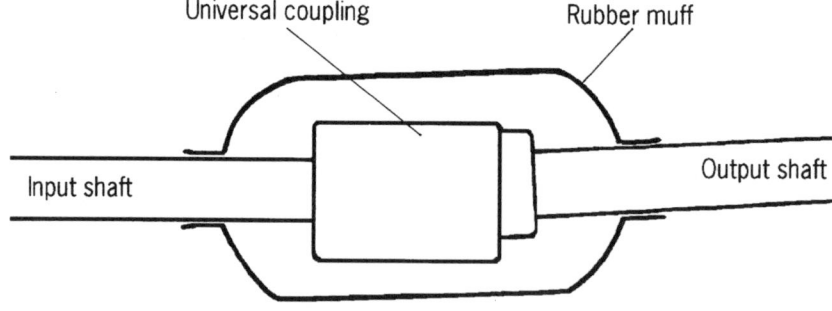

Figure 10.4   Rubber muff as static seal for universal coupling

preventing a flammable lubricant from coming into contact with hot metal.

Flexible semi-static seals are easy to design and manufacture, and can be used in a wide variety of shapes and sizes to suit different applications. However, the life of such a seal depends on the degree of flexing and the number of cycles, and will be reduced where the amplitude and frequency of flexing are high.

10.4 Rotary seals

The majority of seals are used in conjunction with rotating shafts, and there are many different types. For oil-lubricated systems such as crankcases and gearboxes a particular type of rubber lip-seal is most commonly used, to such an extent that this type is commonly called an 'oil seal' (Fig. 10.5). Other types which may be used include simple O-rings, compression packings, and mechanical seals. A special form of lip-seal is used for sealing grease in ball-bearings.

The variety of rotary seals is enormous, and it is intended here only to give a brief description of the most important types used in conjunction with lubricants.

All are dynamic seals, which slide against a second surface, or counterface. It is important to keep the friction between seal and counterface low, and to minimize the wear which takes place. Where a lubricant is being sealed, the friction and wear are both reduced if a thin film of the lubricant can be maintained between seal and counterface. The wear of the counterface is also reduced by making it of a relatively hard material. It is given a very smooth surface by grinding or lapping in order to reduce the wear of the seal and to reduce leakage.

Care is needed to maintain high sealing efficiency. This subject is covered in more detail later.

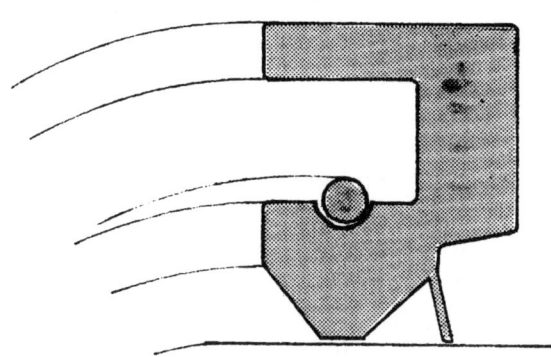

Figure 10.5   Section of a lip seal

(a) *Lip seals*

The mechanism by which a lubricant film is maintained in a lip seal has never been satisfactorily explained. Several different explanations have been given, but none has been generally accepted. The film is almost certainly hydrodynamic, but there is no obvious pressure wedge to generate hydrodynamic pressure. It may be that the asperities on the surfaces create wedges, and thus a micro-elastohydrodynamic lubrication.

The leakage rate depends on the pressure of the lubricant and on the size of the gap between seal lip and counterface, or in other words the thickness of the lubricant film. Just as with a hydrodynamic plain bearing, the ideal condition is where the thinnest possible lubricant film is present. The film thickness decreases if the contact pressure between lip and shaft increases. Too low a contact pressure leads to a thicker film and higher leakage. Too high a contact pressure leads to a film which is too thin and there will be unacceptable friction and wear. Fortunately, the surface tension of the film also tends to prevent leakage, and even a film thick enough to prevent surface contact completely may allow no leakage.

If the lubricant pressure is increased there is a tendency for more leakage through the gap. The lip is, however, designed so that increased lubricant pressure increases the total contact pressure and thus reduces the thickness of the gap. The contribution of lubricant pressure to the total contact pressure is therefore an important factor in maintaining seal efficiency.

If an oil seal is mounted the wrong way round, the lubricant pressure will reduce the total contact pressure, and the seal efficiency will be very much reduced. Many seal housings will permit a seal to be mounted either way around, so it is important to ensure that the seal is fitted correctly. The garter spring should always be on the oil side.

Due to this strong directional effect, a lip seal is not very efficient if used to separate two lubricants, or a lubricant and another liquid. Two separate seals can be used back-to-back, each sealing one of the two lubricants, or a double seal may be used. This situation may arise if a hydraulic fluid must be separated from a gear oil, or a rolling lubricant from the roll support bearing lubricant. However, it is often found that the space available does not allow two conventional oil seals to be fitted, and special seal designs exist to provide the same effect within a limited space.

Simple lip seals are probably limited to a maximum oil pressure of about $100 \, \text{kN/m}^2$ (15 psi) because of the flexibility of the rubber. For higher pressures a steel support ring may be built into the outer part of the seal. Such supported lip seals are usable to almost $1000 \, \text{kN/m}^2$ (145 psi).

If the shaft rotates in only one direction, helical ridges can be incorporated in the bore of the lip to pump oil back towards the pressure side. Although these positive action lip seals will withstand slightly higher oil pressure than conventional lip seals, their main use is where eccentricity or vibration are too high for the conventional types.

The restriction of positive action lip seals to a single direction of rotation is overcome in certain ingenious designs which give positive pumping action with a reversible shaft rotation.

(b) *Compression packings*

These are seals in which a ring or series of rings of a flexible material are mounted in an annular space around the shaft, and are loaded against the shaft surface by some form of axial compression. An example is shown in Fig. 10.6.

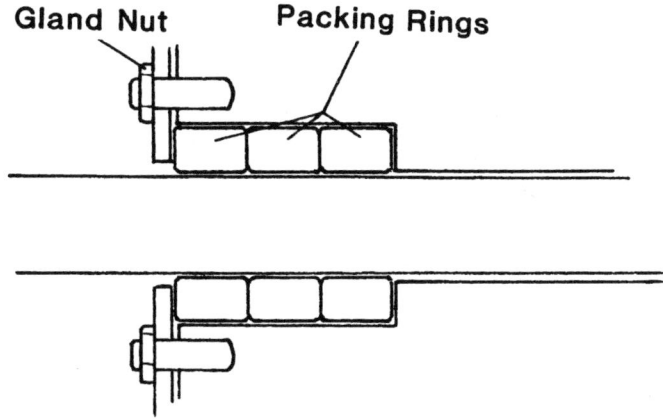

Figure 10.6   Compression packing

Compression packings can withstand much higher pressures than lip seals – up to $30\,\mathrm{MN/m^2}$ (4300 psi) – but are limited to lower rubbing speeds. They are used in high-pressure hydraulic systems, and for many non-lubricant applications.

The simple 0-ring is a similar device, but the loading against the shaft surface is obtained by a small radial interference between the 0-ring and the shaft and housing.

(c) *Bearing seals*

To retain the lubricant in a ball-bearing, and to keep dust or dirt out, it is possible to fit seals to cover the annular gap between the inner and outer races. The axial space available is very small, and the seals used are usually very thin supported lip seals without a garter spring. They depend on the flexibility of the rubber for their contact pressure. They are made rather stiff and may be held in position by thin metal covers.

(d) *Mechanical seals*

In simple terms, the sealing in a mechanical seal takes place on a radial face between two bushes, loaded axially against each other, as shown in Fig. 10.7. One of these bushes is fixed and sealed to the rotating shaft and the other to the housing.

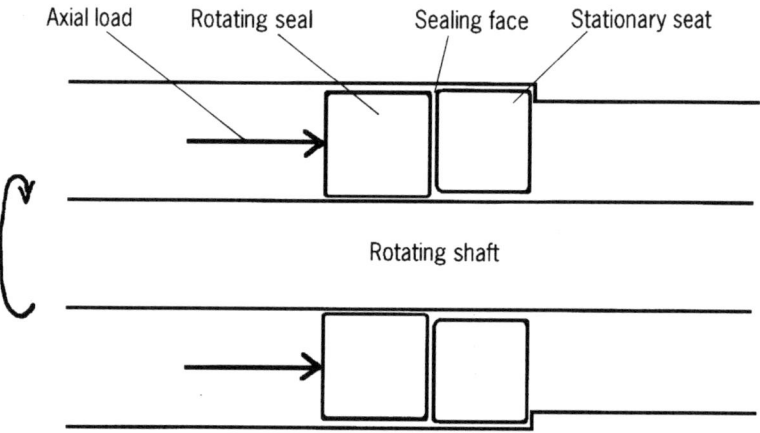

Figure 10.7   Basic design of mechanical seal

Within this simple definition there is an enormous variety of mechanical seals, some of them being of considerable complexity. One of the two bushes is made of, or faced with, a dry bearing material such as graphite or PTFE, and this is normally called the bush. The other is made of steel or some harder material such as tungsten carbide, and is known as the seat. Either the bush or the seat may be fitted to the rotating component. The axial load can be provided by a helical spring, a Belville washer, or hydraulic pressure.

The stationary seal between bush or seat and shaft or housing may be a rubber sleeve, one or more 0-rings, or even a lip seal with a garter spring. In this way, the flexible seals do not also have to cope with sliding motion, and their flexibility no longer limits the maximum sealable pressure. On the other hand, the seal bush and seat surfaces can be made of inflexible materials machined to give very smooth sliding contacts. Being radial instead of circumferential the sealing faces can cope with greater eccentricity.

One example of a mechanical seal is shown in Fig. 10.8, in this case for preventing a process liquid from contaminating a grease-lubricated ball bearing. This particular seal has a graphite sealing bush, able to

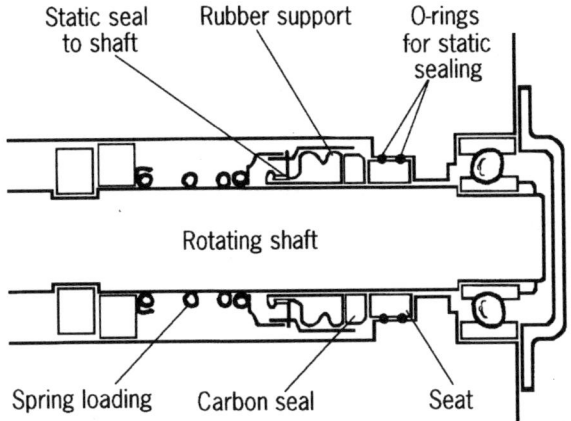

Figure 10.8 Example of a mechanical seal

float radially, but sealed to the shaft by means of a rubber sleeve. The steel seat is sealed into the housing by two 0-rings, and the axial load is provided by a helical spring.

The seal face is normally lubricated by a very thin film of the sealed liquid, but where this liquid is not a good lubricant the sliding properties of the bush material will determine the friction and wear life. The axial spring load maintains a steady contact pressure between bush and seat, and compensates for wear of the seal face. For the same reason the axial positioning is not critical during assembly, within perhaps ±5 percent of the compressible length of the spring from the correct design position.

The pressure of the sealed liquid is often used to provide additional axial contact load at the seal face. In this way an increase in liquid pressure produces increased contact load and maintains effective sealing against the greater pressure. However, with very high liquid pressures the contact pressure may become excessive, causing lubrication failure and excessive wear or surface break-up. For high-pressure applications the seal will therefore be designed so that the hydraulic pressure is used to balance out part of the axial load.

Unbalanced mechanical seals can be used to about 1 MN/m^2 (145 psi) while balanced types can be used to 10 MN/m^2 or even, with special designs, to over 20 MN/m^2 (2900 psi).

(e) Clearance seals

All the types of rubbing seal described so far have several disadvantages. They are complicated and expensive, critical in fitting, wear out in service and generate friction. If the liquid which is being sealed is not flooding the seal location, but is only reaching it by splashing, or creeping along the shaft, it may be possible to provide adequate sealing by means of a non-rubbing, or clearance seal. The most common types are labyrinths and bush seals. These are depicted in Fig. 10.9.

The floating bush seal, Fig. 10.9 (c) is very widely used. It has the advantage that shaft eccentricity is compensated for by radial movement of the bush, so that a smaller bush clearance can be used.

If the direction of shaft rotation is constant, a helical labyrinth can be mounted on the shaft so as to pump the liquid back by a screw-feed mechanism. This is called a viscoseal.

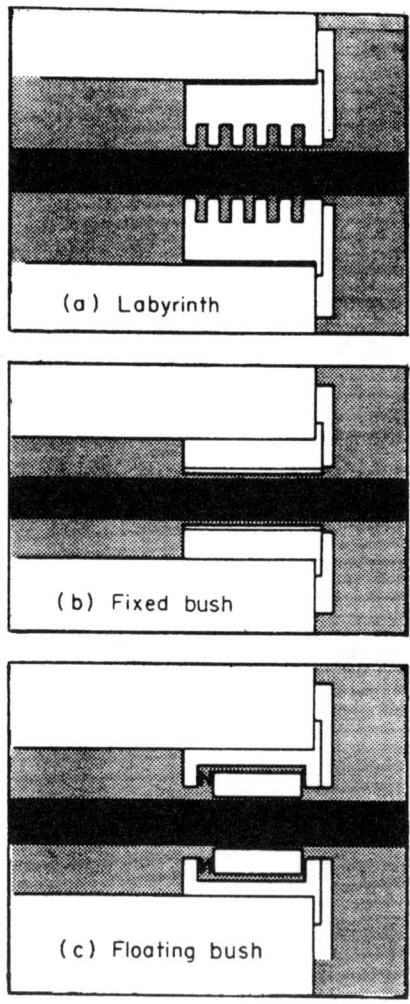

Figure 10.9 Three types of clearance seal

(f) *Flinger rings*

Where the sealed liquid reaches the seal location mainly by flow along the shaft, much of it will be flung off by centrifugal action before reaching the seal. This effect can be improved by mounting a ring on the shaft. The centrifugal force is proportional to the square of the ring diameter, so that it is an advantage to use a ring of the highest convenient diameter. A ring with a diameter twice that of the shaft will have four times the throwing effect. In addition, however, any liquid which flows on past the flinger ring must flow inward against the centrifugal effect. This will further reduce the amount of liquid passing the ring location.

It will sometimes be possible to mount a flinger ring just inside the casing, so that the oil flung off from the ring returns immediately to the sump. More often the ring will be mounted inside a housing, and it must then be co-located with drain grooves in the housing. Several flinger rings and drain grooves may be fitted. These can give excellent control of the liquid, even where there is extensive splashing and spraying taking place, as in a high-speed gearbox or crankcase.

10.5   Sealing reciprocating shafts

Many of the seals already described can be used with reciprocating motion of a shaft. For small amplitudes, bellows or even diaphragms may be suitable. For low speeds with good lubricants 0-rings, packed glands, and even lip seals can be used.

There are, in addition, several types specially designed for use with axial movement. A few of the most important are U-rings, chevron packings, and piston rings.

(a) *U-Rings*

Figure 10.10 shows a typical U-ring in position as the seal for a hydraulic actuator rod. It consists of a rubber ring with a U-shaped cross-section. The loading of the lip against the rod surface is given by the deformation of the ring shape by compression between rod and housing.

The ring should preferably be installed so that the rod moves away from the open side of the 'U' when that side is under pressure. Thus in Fig. 10.10 the rod should be moving to the right when the hydraulic fluid to the right of the seal is at high pressure, and to the left when the fluid pressure is reduced.

All rubber ring seals in reciprocating systems must be installed so that the extent to which the rubber protrudes from the housing (i.e., the

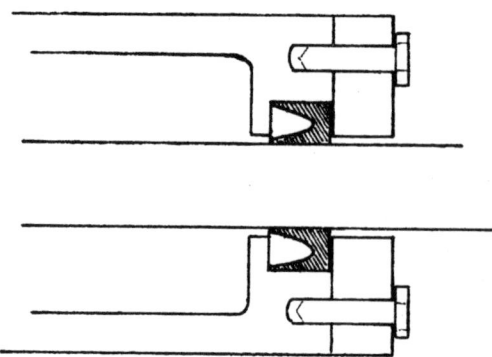

Figure 10.10   U-ring seal for a hydraulic actuator rod

clearance between shaft and housing bore) is small compared with the axial thickness of the ring.

(b)   *Chevron packings*

One common form of compression packing is the chevron, in which the cross-section of the compressible ring is in the form of a broad 'V'. This form is particularly suitable for reciprocating shafts. Chevron packings can be used singly or in multiple sets. When used as rotary seals they are compressed mechanically. For reciprocating use they are mounted between shaped metal rings with little or no mechanical pressurization. The necessary pressurization is given by the sealed liquid.

It is sometimes considered desirable to provide some means of axial adjustment of the seal housing. In normal operation this would be used only to eliminate play during assembly. If excessive leakage occurs later, due to seal or rod damage, it may be useful as a strictly interim measure to tighten up the adjustment and compress the chevron to give a tighter fit on the rod and reduce leakage pending a proper repair.

(c)   *Piston rings*

The most familiar type of piston ring seal is probably the split metal ring used in an automotive piston engine, in which the ring is split axially at one point. The ring is mounted in a square section circumferential groove in the piston under slight compression, so that it expands tightly against the cylinder wall. The split in the ring leaves a very small gap in the seal, but this is acceptable – especially since three or more rings are normally used in series.

The chief value of the all-metal piston ring is that it withstands the high temperatures and oxidizing conditions in the cylinder of an internal combustion engine. They can only be used with lubrication, although a solid lubricant such as PTFE or a molybdenum disulphide composite has been used with some success, in the form of an additional ring.

Where the temperature is lower, especially if liquid lubrication is not possible, similar piston rings made of reinforced PTFE, carbon, graphite, or various lubricating composites can be used. Rubber U-rings and chevrons can also be used as well as many modified forms, such as double U-rings, W-rings, and lobed 0-rings.

10.6 Seal materials

In selecting seal materials, their compatibility with the lubricant must be taken into account in accordance with Table 4.3. The maximum temperature limit will also be important. Natural rubber can be used to about 80 °C, nitrile to 120 °C, and viton and silicone rubbers to 200 °C.

10.7 Handling and fitting

Probably the most important single factor in seal performance is the smoothness and conformity of the seal and counterface surfaces. Any roughness or scoring of either will provide leakage paths and destroy the effectiveness of the seal.

Flexible seal materials are generally soft and easily damaged, and they should be stored and handled carefully. For assembly it will often be helpful to moisten slightly with the type of lubricant which is present in the system.

Lip seals, 0-rings, and other rubber rings often have to be assembled onto a shaft having locating grooves, keyways, ports, threads, and so on. The rubber lip can easily be damaged by any sharp discontinuity.

Wherever possible, grooves, keyways, and ports should be slightly radiused before seals are fitted, and threads should preferably be cut below the main shaft diameter. A thin metal or hard plastic sleeve can often be fitted over a thread or other discontinuity to prevent damage when a seal must be slid past them.

Chapter 11

Lubricant Testing and Specifications

11.1 The object of testing

The most important factor in testing is to match the tests to the objective of testing. To do this it is necessary to be clear about just what the purpose of the tests is.

Testing of a lubricant, or of anything else, may be done for a number of different reasons and the following are some examples:

(a)   to find out if it will work well in a particular piece of equipment;
(b)   to find out if it will survive in particular conditions, such as heat or humidity;
(c)   to ensure that it is safe to use, and not toxic or highly flammable;
(d)   to ensure that it meets a specification;
(e)   to find out if it is the same type of oil as a previous shipment;
(f)   for legal purposes, for example to find out if it meets a flammability limit, or falls into a particular category for excise duty.

The tests which are used for lubricants fall into three main categories.

(a)   *Functional tests*
   These are designed to show how the lubricant performs its function as a lubricant, or one of its secondary functions as a hydraulic fluid, coolant, or corrosion preventive. They usually involve simulating, to some extent, the conditions under which the lubricant will be expected to operate.

(b)   *Chemical tests*
   These involve measuring some chemical property of the lubricant, such as acidity, or the concentration of some chemical substance which is present in it, such as calcium or naphthenes.

(c) *Physical tests*

Physical properties may be measured either because they are themselves important, as with viscosity, or because they give information about some other important aspect, as when colour indicates the state of a used lubricant.

The effort and cost involved in testing is, of course, important. In general, functional tests are most directly useful in predicting how a lubricant will perform in service, but they are also the most expensive. Chemical tests are very useful because, if enough is known about the chemical nature of the lubricant, this will go a long way towards predicting how it will perform in service. Chemical tests tend to be much cheaper than functional tests, but more expensive than physical tests.

Physical tests, although the cheapest, are also usually the least valuable in predicting service performance. An exception is viscosity measurement, which is relatively inexpensive but extremely important in that it directly predicts hydrodynamic performance. On the other hand, even if the material tested has a viscosity of, say, 20 cSt, and this is ideal for the application, the information is not much use if it is not known whether it is a lubricant, an adhesive, or sulphuric acid. In other words, physical test results generally have to be used in conjunction with information from other sources.

The usefulness of a test is increased considerably if the result can be compared directly with results obtained by others. For this reason there has been a great development of standardized test methods, especially during the last fifty years.

11.2 Functional tests

If a bearing, or a machine, is completely conventional, and operating well within its limits, the selection of a lubricant for it may be quite straightforward, and no testing will be needed. On the other hand, if the design is unconventional, or if the load, speed, temperature, or some other factor is unusual, then testing will be essential.

Prediction of lubricant and bearing behaviour in unusual conditions is not highly reliable. There is then only one way to make sure that the lubricant will perform satisfactorily in a particular piece of equipment and under particular service conditions. That is to carry out a service trial in which the lubricant is tested in the actual service equipment under actual service conditions.

Unfortunately, there are several factors which make service trials unsatisfactory.

(a) *Duration of the trial*

    If the life expected of the lubricant in service is a year, or five years, then the service trial will have to last longer than that, to make sure that there is a margin available over the minimum life required. In most cases it would not be possible to wait that long for a result.

(b) *Insufficient control*

    It is often not possible in a service trial to cover all the possible environmental conditions which may be encountered eventually, from arctic cold to tropical heat, or from jungle humidity to desert dryness.

(c) *Difficulty of monitoring*

    It is usually not possible in a service trial to measure important factors such as friction, bearing temperature, and wear rate. If a bearing failure occurs, the damage may be too great to show whether the lubricant or some other quite accidental factor caused the failure.

(d) *Cost*

    A service trial will often be expensive, so that although one will eventually be needed, some cheaper form of sorting test may be preferable initially. On the other hand, a well-controlled laboratory test can sometimes be more expensive than a service trial.

    An alternative to a service trial is a laboratory test based on the actual service equipment. This has many advantages.

(a) The operating conditions can be controlled and varied over a wide range, if necessary covering every set of conditions which might be met in service.

(b) Special instruments can be fitted to measure operating performance, such as thermocouples to measure oil and bearing temperature, accelerometers to monitor vibration, and torque meters to measure friction.

(c) The results may be obtained sooner by accelerating the test, although this must be done with great care. An increase in load, for example, may wreck a bearing which would operate perfectly well under the actual service conditions. Increasing the speed may also wreck the system or, on the other hand, may help it to survive by taking the bearing conditions from boundary to hydrodynamic. One way which is generally safe is to reduce the amount of idle time. For example, in service a generator might operate only twice a day for a total of one hour. In a laboratory test it might be run thirty times a day for a total of fifteen hours without changing the operating conditions.

Figure 11.1 shows a bank of compressors set up for laboratory testing of lubricants, with additional gauges and filters fitted to improve monitoring and control. The most important class of laboratory lubricant tests based on actual service equipment comprises the many engine tests used to develop and evaluate automotive engine oils.

There are many situations in which it is impossible to use the service equipment for testing, either in the field or in the laboratory. This may be because it is too expensive, or because only one example will be built, as with a ship or a spacecraft, or because the design has not yet been finalized. Under these circumstances laboratory test rigs must be used.

When a piece of service equipment is used for testing, there is always a good simulation of the shape and the operating conditions. The simulation is never as good with a test rig, but efforts must always be made to see that the following important factors are imitated as accurately as possible.

- Contact shape
- Bearing size
- Bearing speed
- Contact pressure
- Lubricant flow
- Temperature
- Environment

Some laboratory test rigs are built to simulate the bearing or other components for which the tests are being made, and it may then be

Figure 11.1 Compressors set up for oil testing

possible to reproduce all the above factors. Often, however, a general purpose test rig will be used. It should be capable of imitating as many of the important factors as possible, especially the speed, pressure, and temperature. Since the simulation is not complete, tests should also be carried out over a range of speeds, pressures, and temperatures, to make sure that there is a safety factor available in the lubricant performance.

Many general purpose laboratory test rigs are commercially available. Figure 11.2 shows one which can be used to test lubricants in a wide variety of different journal bearing types.

The discussion so far has concentrated on lubricant performance testing from the user's point of view. The problem for the lubricant manufacturer is different. It is impossible for an oil to be tested in every type of equipment in which it might be used, although in fact a major oil company might test a new engine oil in thirty different types of engine. At the early development stage, however, test rigs are needed which will test small quantities of oil or grease in such a way that one can be reasonably confident of applying the results to a wide range of service uses. Table 11.1 lists a number of the most common test rigs of this type.

Perhaps the best-known of all lubricant test machines is the Seta-Shell four-ball machine (Fig. 11.3). This is often criticized because it does not simulate any real bearing geometry. However, the results obtained with it have been used successfully in the development of hundreds of different lubricants.

Figure 11.2 Coturnix journal bearing test machine

Table 11.1   Some common lubricant test machines

Test rig	Test method
Seta-Shell four-ball lubricant test machine	IP239, ASTM D–2596
Timken wear and lubricant testing machine	IP240, ASTM D–2509
Falex lubricant tester	ASTM D–3233
IAE gear machine	IP166
FZG gear machine	IP334
Ryder gear test machine	ASTM D–1947
Rolling bearing test machine	IP168
Alpha LFW-1 friction and wear testing machine	ASTM D–2714

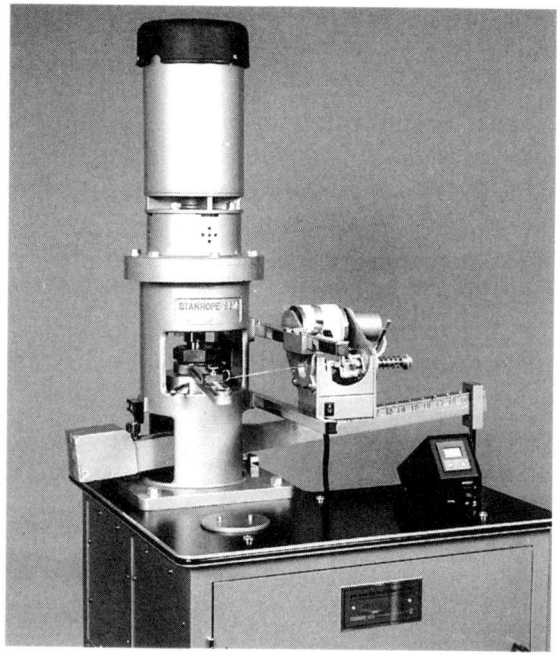

Figure 11.3   Seta-shell four-ball machine

The results obtained from such machines are often quoted in specifications. These are commonly used in purchase agreements, so the test conditions must be tightly controlled. Many of these machines have been the basis of test methods standardized by the British Institute of Petroleum (IP), the American Society for Testing and Materials (ASTM) and other standardization organizations. Some of these standard methods are listed in Table 11.1.

There is yet another class of functional tests for lubricants in which it is not the lubricating performance but some other function which is tested. These are also often standardized methods. Several of the more widely recognized ones are listed in Table 11.2. They include tests for corrosion, flammability, emulsification, and so on. On the whole, no attempt is made with such tests to simulate the conditions of any one particular application. The basic assumption is that if lubricant A is better than lubricant B in the test, then A will also be better than B in service. Where the tests have been nationally standardized, it can reasonably be taken that this assumption is likely to be correct.

Table 11.2   Other functional tests for lubricants

Test	Method numbers
Aqueous cutting fluid-corrosion of cast iron	IP125
Rust-preventing characteristics of steam-turbine oil	IP135, ASTM D–665
Foaming characteristics of lubricating oils	IP146, ASTM D–892
Evaporation loss of lubricating greases	IP183, ASTM D–972
Water washout characteristics of lubricating grease	IP215, ASTM D–1264
Dynamic anti-rust test of lubricating greases	IP220
Seal compatibility index of petroleum oils	IP278
Rust prevention characteristics of metal working fluids	IP287
Stability of water-in-oil emulsions	IP290
Water separability of petroleum oils and synthetic fluids	ASTM D–1401
Frothing characteristics of water mix metal working fluids	IP312
Leakage tendencies of automotive wheel bearing greases	ASTM D–1263

Thus functional tests may vary from perfect simulation in a service trial to complete non-simulation in a standardized laboratory test. All of them can give valuable information, provided that the results are interpreted in relation to results obtained from other sources, and that the differences between the test conditions and those which will arise in service are kept in mind.

11.3   Chemical and physical tests

Even if only nationally or internationally recognized methods are considered, there are probably hundreds of different chemical and physical tests which can be applied to lubricants. It would be a waste of space to attempt to describe all of them, and anyone faced with the problem of devising lubricant test programmes will be well advised to obtain the IP and ASTM volumes listed in the Bibliography. It will be enough here to describe some of the more important types of method, and those which relate to other parts of this book.

An excellent guide to the use of tests for lubricants is the 'Lubricants' chapter of *Criteria for Quality of Petroleum Products* (Institute of Petroleum, 1961), although some of the details are now obsolete.

(a) *Oxidation tests*

Oxygen does not dissolve readily in oils, and most oxidation is caused by free oxygen in the form of a gas. The rate of oxidation depends on the extent to which the oxygen comes into contact with the oil, in other words the amount of oxygen present and how much mixing takes place, as well as on the temperature and any other substances present which may catalyse (increase) or inhibit (decrease) the rate of oxidation. As a result, many different oxidation tests have been developed to simulate different service conditions. Some are listed in Table 11.3.

The results of one test cannot be compared directly with another, so users need to choose the test which is most appropriate for their own application. For example, IP229 is most suitable for steam turbine oils, and IP 176 is most suitable for engine oils. IP229, IP280, and IP328 can be applied to used oils to determine their remaining oxidation service life.

Table 11.3 Standardized oxidation tests for lubricants

Test	Method numbers
Oxidation characteristics of lubricating oil	IP43
Oxidation stability of lubricating greases (bomb method)	IP142, ASTM D–942
Oxidation characteristics of inhibited mineral oils	ASTM D–943
Oil oxidation and bearing corrosion (Petter SI engine test)	IP176
Oxidation stability of steam turbine oils (rotating bomb)	IP229, ASTM D–2272
Oxidation stability of inhibited mineral turbine oils	IP280
Oxidation stability of straight mineral oil	IP307
Oxidation stability of mineral insulating oil	IP307
Oxidation stability of mineral turbine oils during use	IP328
Oxidation stability of inhibited mineral insulating oils	IP335
Oxidation characteristics of extreme pressure oils	ASTM D–2893
Oxidation induction time-lubricating greases	ASTM D–5483
Sludging tendency of inhibited mineral oils	ASTM D–4310
Oxidation stability of gasoline engine oils	ASTM D–4742

(b) *Thermal stability*

Thermal stability is usually less important in practice than oxidation stability, and does not depend as strongly on the conditions of use. There is therefore less need for a wide variety of different tests for thermal stability. Those listed in Table 11.4 fall basically into three groups.

Table 11.4 Some tests for thermal deterioration

Test	Method numbers	
Thermal decomposition		
Thermal stability of water mix cutting fluid	IP311	
Thermal stability of hydraulic fluids		ASTM D–2160
Thermal breakdown of greases		
Dropping point of lubricating grease	IP132,	ASTM D–566
Heat stability of calcium-base greases	IP180	(obsolete)
Dropping point of lubricating grease, wide temperature range		ASTM D–2265
Loss of volatile components		
Evaporation loss of lubricating grease	IP183	
Evaporation loss of lubricating greases and oils		ASTM D–972
Evaporation loss, lubricating greases, wide temperature range		ASTM D–2595

IP132, IP180, ASTM D–566, and ASTM D–2265 are concerned with breakdown of the structure of greases at high temperature. IP183, ASTM D–972, and ASTM D–2595 measure the extent to which volatile components of oils or greases evaporate away at high temperature.

Only IP311 and ASTM D–2160 are designed to show the tendency of an oil to break down chemically as the temperature is raised. In both cases the tests are used to assess the effect of temperature in a situation where little or no oxygen is present.

(c) *Flammability*

Like oxidation, flammability of a lubricant depends very much on the conditions present, and not simply on the temperature. As a result a variety of different tests are used, some of which are listed in Tables 11.5 and 11.6, but the results cannot be compared from one test to another. The effects of different conditions are discussed more fully in Chapter 14, together with some of the relevant tests.

Due to the great effect of different test conditions on ignition behaviour, it has been difficult to obtain agreement on the selection of standard tests. This is illustrated by Table 11.6, which lists eight

Table 11.5 Some flammability tests

Test	Method numbers
Autogenous ignition temperature	ASTM D–2155
Reaction threshold temperature	ASTM D–2883
Mist spray flammability	ASTM D–3119
Linear flame propagation rate	ASTM D–5306
Autoignition determination of liquid chemicals	ASTM E 659
Hot manifold test	AMS 3150c
Flammability spray test for hydraulic fluids	BS DD61

Table 11.6 Commonly used flash point tests

Test	Method numbers
Petroleum (Consolidation) Act 1928, method (Abel apparatus)	IP33 (obsolete)
Closed flash point (Pensky-Martens)	IP34, ASTM D–93
Open flash and fire point (Pensky-Martens)	IP35
Open flash and fire point (Cleveland method)	IP36, ASTM D–92
Flash point (Abel closed cup method)	IP170
Flash point (Setaflash closed tester)	IP303, ASTM D–3828
Flash test closed cup equilibrium method	IP304
Flash point by Tag closed tester	ASTM D–56

different widely used flash point tests. The general principle of all of them is the same, in that a sample of the test liquid is heated and a flame is brought near its surface. The lowest liquid temperature at which the vapour above the surface ignites is the flash point. The many methods differ in the size of container, whether it is open or closed, the rate of heating, and so on, all of which can affect the flash point.

Flash points are the most commonly used flammability tests, and are usually quoted for any liquid lubricant. They are also used to define flammability for certain legal purposes. In spite of this, however, their precision is poor.

Flash points of mineral lubricating oils are generally above 160 °C, but may be as low as 80 °C for light lubricating and hydraulic oils. The flash point test is a useful test for fuel contamination of engine oils, as even 5 percent of a fuel in an engine oil will reduce the flash point considerably.

(d) Composition

There are over forty Institute of Petroleum standard tests and a similar number of ASTM tests which can be used to obtain information about the composition of lubricants. These are in addition to those used mainly to analyse contaminants.

One important group are the tests for acidity and alkalinity, listed in Table 11.7. Alkalinity, or base number, is an indication of the presence of alkalyne additives, such as sodium or calcium compounds. Base number tests may be carried out on new lubricants as a quality control method, or on used lubricants, where they indicate the amount of additive still available.

Table 11.7   Tests for acidity and alkalinity

Test	Method numbers
Determination of acidity – neutralization value	IP1
Acidity and alkalinity of lubricating grease	IP37
Saponification number – titration method	IP136, ASTM D–94
Neutralization number by colour indicator titration	IP139, ASTM D–974
Neutralization number (potentiometric titration)	IP177, ASTM D–664
Inorganic acidity of petroleum products	IP182
Total base number potentiometric – perchloric acid titration method	IP276, ASTM D–2896
Total acid number by semi-micro colour indicator titration	ASTM D–3339
Base number of petroleum products (potentiometric titration)	ASTM D–4739

Acidity is also used as a quality control test to monitor the purity of an unused lubricant. A typical acid number for an unused lubricant is less than 0.3 mg KOH/g. Acidity is also a valuable check on a used lubricant, as it indicates the degree of oxidation which has taken place.

There are also many other tests for composition, and some examples are listed in Table 11.8.

More general information about the composition of the base oil is easily obtained by the technique called infra-red spectroscopy. This is a cheap technique requiring only a tiny sample, and distinguishes all the types of synthetic oil and the main composition types in mineral oils, but it requires specialist knowledge to carry it out. Some examples of infra-red spectra of different base oils are shown in Fig. 11.4.

(e)   Contamination

The subject of contamination analysis is discussed in more detail in Chapter 11, but some of the more important standard test methods for contaminants are listed in Table 11.9. In addition, of course, any composition test is a test for contaminants if it is used to detect something which was not originally present. Thus IP117 is used to

Table 11.8 Some tests for lubricant composition

Test	Method numbers	
Sulphur (bomb method)	IP61,	ASTM D–129
Calcium content of lubricating oil	IP111	
Zinc content of lubricating oil	IP117,	
Chlorine in new and used lubricants	IP118,	ASTM D–1317
Asphaltenes (heptane insolubles)	IP143	
Phosphorus content of petroleum products (flask method)	IP245	
Total metals in unused lubricating oils and additives	IP339	
Additive elements – lubricating oils (ICP–AAS)		ASTM D–4951
Additive/wear elements – used lubricating oils		ASTM D–5185
Phosphorus in lubricating oils and additives		ASTM D–1091
Lubricant additive components (X-ray fluorescence)		ASTM D–4927
Total nitrogen in lubricating oils (Kjeldahl method)		ASTM D–3228
Lithium and sodium in lubricating greases		ASTM D–3340

analyse zinc-containing additives, but could equally detect zinc contaminant in an oil.

Similarly, flash point determinations will indicate the presence of low-flash point contaminants (fuels or solvents) in lubricating oils.

(f) Viscosity and consistency

A number of test methods for viscosity and consistency are listed in Table 11.10. The preferred test method for viscosity of an oil is the capillary viscometer method standardized in IP71, ASTM D–445. The method is based on the measurement of the time taken for a known volume of oil to flow under gravity through a calibrated glass capillary viscometer. A typical capillary viscometer is shown in Fig. 11.5. This gives the kinematic viscosity in cSt. The dynamic viscosity in cP is obtained by multiplying the kinematic viscosity by the density of the oil in grams per cubic centimetre at the same temperature.

The capillary viscometer is not suitable for use with extremely viscous liquids. For this reason oil viscosity at very low temperatures is measured by rotary systems in IP267 and ASTM D–2602. The lowest temperature at which the oil will flow at all is indicated by the pour point method, IP15/ASTM D–97.

The consistency of a grease is normally measured by the cone penetration method (described earlier) of IP50/ASTM D–217, or if smaller samples must be used, by the smaller scale versions of IP310/

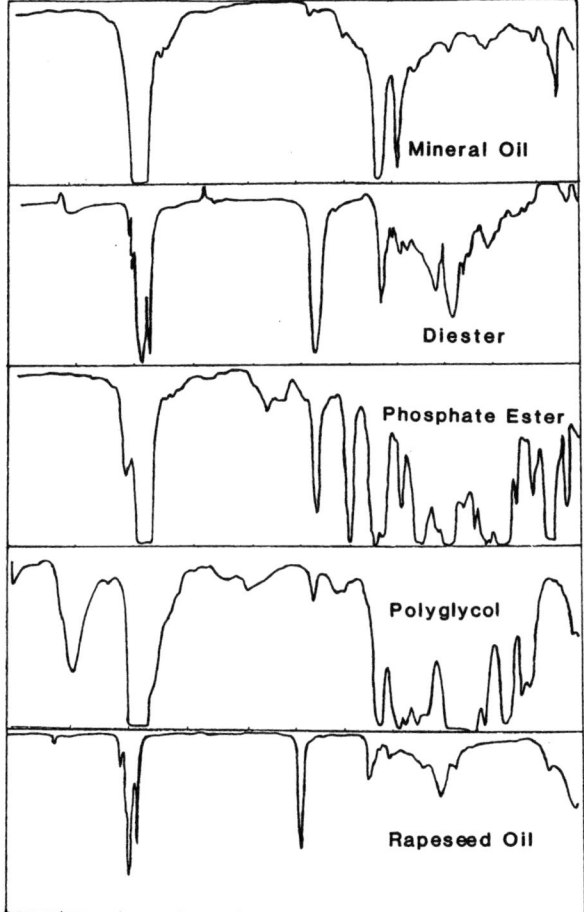

Figure 11.4   Infra-red spectra of different base oils

Table 11.9   Some tests for contaminants in lubricants

Test	Method numbers
Gasoline engine crankcase oil fuel dilution	IP23, ASTM D–322
Water content of petroleum products (Dean & Stark)	IP74, ASTM D–95
Foreign particulate matter in lubricating grease	IP134
Sizing and counting particles from hydraulic fluids	IP275 (obsolete)
Total solids in used engine oils	IP316
Deleterious particles in lubricating grease	ASTM D–1404
Water in liquid petroleum products (Karl Fischer)	ASTM D–1744
Glycol-base antifreeze in used lubricating oils	ASTM D–2982
Contaminants in used lubricating oils (ICP-AES)	ASTM D–5185

Table 11.10 Tests for viscosity and consistency

Test	Method numbers
Oils	
Pour point of petroleum oils	IP15, ASTM D–97
Kinematic viscosity and calculation of dynamic viscosity	IP71, ASTM D–445
Calculation of viscosity index from kinematic viscosity	IP226, ASTM D–2270
Low temperature viscosity of automotive fluids	IP267
Apparent viscosity of motor oils at low temperatures	ASTM D–2602
Apparent viscosity (cold cranking simulator)	ASTM D–5293
Greases	
Cone penetration of lubricating grease	IP50, ASTM D–217
Dropping point of lubricating grease	IP132, ASTM D–566
Low temperature torque of lubricating grease	IP186, ASTM D–1478
Cone penetration of grease (small scale)	IP310, ASTM D–1403
Apparent viscosity of lubricating greases	ASTM D–1092

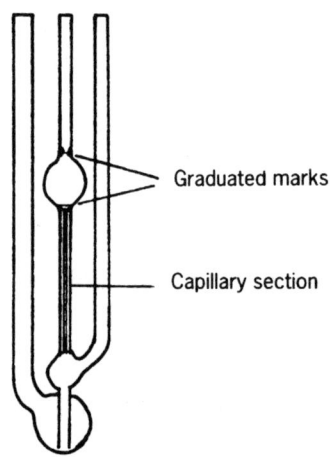

Graduated marks

Capillary section

Figure 11.5 Suspended level capillary viscometer

ASTM D–1403. The temperature at which a grease begins to flow under gravity is given by the dropping point method IP132/ASTM D–566. The apparent viscosity of a grease when it is made to flow is indicated by ASTM D–1092. Due to the difficulty of predicting grease behaviour, the critical problem of starting greased bearings at low temperatures is assessed directly by the low-temperature torque tests of IP186 and ASTM D–1478. Unfortunately the precision of these tests is poor.

(g) *Other physical tests*

The density or relative density (specific gravity) has some importance in itself, as for example in converting kinematic to dynamic viscosity, but is more often measured as a quality control technique in the manufacture or sale of lubricants. It is usually measured by means of a hydrometer (IP160, ASTM D–1298) or a pycnometer (IP189, IP190).

The colour of a lubricant is also assessed as a quality control technique, and one widely standardized method is IP196, ASTM D–1500, 'ASTM Colour of Petroleum Products'.

11.4 Standards and specifications

Lubricants are generally bought or sold by the manufacturers' brand names, such as BP Visco 2000, Shell Rimula X15/40, Esso Dortan 12, or Mobilgrease 28. Once a user has established that a particular brand is satisfactory for his application, the brand name may be a sufficient guarantee of performance. On the other hand, a brand name gives no information on performance to a potential user seeing it for the first time, and it needs to be supported by a document giving properties and performance figures.

Such a document is a specification, although it may be called a data sheet or a standard. In general, data sheets are issued by lubricant suppliers to describe typical properties of one or more of their products. A typical data sheet might list the properties shown in Table 11.11.

A supplier's data sheet may not give the user any direct information showing that the lubricant is suitable for a particular application, but it does two useful things. The first is to indicate to the user the typical properties which the supplier tries to maintain for that brand, and thus give the user some yardstick by which to assess the lubricant. The second is that if data sheets for two different brands list very similar properties, the user can have increased confidence that it is possible to change safely from one brand to another.

Table 11.11 Typical data sheet properties

| | Oil type | | |
	No 1	No 2	No 3
Viscosity: cSt at 40 °C	75	165	240
cSt at 100 °C	8.9	14.8	18.4
Viscosity index	95	95	95
Pour point, °C	−20	−10	25
Acidity, mg KOH/g, max	0.2	0.2	0.25
Flash Point, closed cup, °C	195	210	215
Specific gravity	0.887	0.894	0.900

Remember, however that similar data sheets give *no* guarantee that the two brands concerned can be interchanged safely, especially in some application not covered by the data sheets. For example, two hydraulic fluids from different manufacturers may have almost identical data sheets. However, if no data on cleanliness is given, one of them may work well in a system while the other, being dirtier, causes plugging of valves and complete failure.

To avoid this problem, major users will often draw up their own specifications which cover the performance aspects which are important in their own applications. In the example quoted above, the user might specify limits on the solid particle contamination, and in this way avoid the risk of damage from a dirty fluid.

Technically, a specification differs from a data sheet in that it lays down limits for various properties or test results, as shown in Table 11.12. A specification is often intended to be the basis for a purchasing contract, and the specification limits then become contractual obligations. For this reason, it may not be possible for a small user to persuade a major supplier to supply a lubricant to meet the user's specification.

Major users, such as Government Departments or large transport undertakings, usually represent a big enough market for them to be able to insist that lubricant supplies meet their purchase specifications. Common examples are the defence departments of most countries, and national railways and utility companies.

In other cases major groups of users, or joint groups of suppliers and users, may produce a specification covering joint needs. Examples of such specifications are those of the ASTM and SAE.

The small user may be able to take advantage of such specifications in one of two ways. If an application is similar to one for which a major specification exists, then it may be possible to buy a lubricant certified as meeting that specification. Alternatively, if a lubricant supplier advertises a branded product as meeting the specification, it can be purchased with confidence that it does so.

Table 11.12 Typical specification requirements

Test	Limits	Methods
Viscosity, kinematic, cSt at 40 °C	60–75	IP71/87*
Viscosity, kinematic, cSt at 100 °C	8.3–8.5	IP71/87
Flash point, closed, °C	Min 190	IP34/88
Pour point, °C	Max −10	IP15/94
Total acid number, mg KOH/g	Max 0.1	IP177/83
Copper corrosion, classification	Max 1	IP154/93

*The second number indicates the date of most recent revision. This date can often be omitted in general use, but in specifications it should always be quoted.

When a specification becomes so widely used that it comes to represent the normal level of quality or performance expected from a particular class of product, it may become a standard. Standards are generally established at a national or international level.

The older standards-making organizations are all national, the oldest of all being the British Standards Institution. There are now over eighty national standards organizations, of which the more important include the following:

British Standards Institution	BSI	Britain
American National Standards Institute	ANSI	United States of America
Deutsche Institut für Normalisation	DIN	Germany
Association Francaise de Normalisation	AFNOR	France
Gosudarstvyeny Standart	GOST	Russia

The increases in international trade and communications in recent years have brought a need for standardization on an international basis. In the field of lubricant standards, the first important international organization was the Conseil Européen de Normalisation (CEN), but the most important is now the International Standards Organisation (ISO).

In the United Kingdom many of the petroleum test methods developed by the Institute of Petroleum have been adopted and published by the British Standards Institution in the BS2000 series, retaining the IP Method number as a subsidiary part number. For example IP50 *Cone Penetration of Lubricating Grease* is published by BSI as BS2000: Part 50.

Many ISO standard test methods are also based on original IP or ASTM test methods, and IP50 is also internationally standardized as ISO 2137.

Standards may be of many different types. Some are typical specifications, and BS4475 is a specification for straight mineral lubricating oils. Others cover test methods such as those in the BS2000 series. Yet others are classifications, such as BS4231 *Classification for Viscosity Grades of Industrial Liquid Lubricants* or guides to recommended practice, such as BS2000: Part O: Addendum 2 *Significance and Usage of IP Precision Data*.

As far as lubricant tests are concerned, the most important standardizing bodies are still the Institute of Petroleum and the American Society for Testing and Materials. Eighty important methods are published jointly by both bodies, such as IP34/ASTM D93 *Closed Flash Point* (Pensky-Martens). Sixty IP methods, including thirty-five IP/ASTM joint methods, are published as British Standards in the BS2000 series. Forty IP/ASTM Joint Methods are also published as ISO methods.

This broad standardization of test methods leads to a wide acceptance of specifications for lubricants, and thus to wide acceptability of performance standards. There are also considerable advantages to the use of standardized lubricants, since standardization over a wide geographical area is likely to ensure more ready availability.

Basically, the drawing up of a specification is simply a matter of deciding what composition limits and what tests are necessary to make sure that the product will be satisfactory in use. This will be illustrated later in Section 11.6.

11.5 Automotive engine oil specifications

A brief summary of the requirements, composition and properties of automotive engine oils was given in Chapter 4. It is a curious fact that, although such oils are very similar, their specification situation is complicated and confused. This is probably because the consumption of vehicle engine oils is greater than that of all other lubricants put together. They represent the greatest source of income for the suppliers, and considerable financial involvement for the users.

With the introduction of detergent oils certain engine manufacturers, notably General Motors and the Caterpillar Tractor Company, developed their own engine tests to assess diesel engine lubricants. A number of those tests were incorporated in the first US Army specification 2–104B in 1942. This was replaced by the military specification MIL–O–2104 in 1950, and subsequently MIL–L–2104A. The latter, in turn, went through several stages of amendment, and the latest version at the time of writing is MIL–L–2104F.

In 1951 the British Ministry of Supply introduced a specification DEF–2101 which was technically equivalent to MIL–O–2104. This was also amended from time to time, and the last version was DEF–2101D Am. 1. DEF–2101 has now been supplanted by a variety of DEF STAN and TS specifications. The resulting complexity and unfamiliarity has probably been one cause of the virtual disappearance of these specifications for non-military use.

The original Caterpillar L1 engine test was progressively increased in severity, leading to Series 1, Series 2, and Series 3 oils. The last of these was matched by the US Army which introduced a specification MIL–L–45199A for Series 3 oils. Similarly the British DEF–2101B specification included a procedure for testing oils with higher sulphur fuel (i.e., comparable with Series 1 and 2 oils) and these oils were known as Supplement 1 oils.

All these terms, MIL–L–2104, MIL–L–45199, DEF–2101, with all their later amendments, Series 3 and Supplement 1, may be found in oil descriptions, although strictly speaking the Government specifications only apply to oils they approve for their own use.

In parallel with these developments, the American Petroleum Institute (API) in 1952 introduced a classification system for diesel engine oils which was based on the severity of engine service. This led in 1969 to cooperation with ASTM and SAE to produce a classification system which is now in widespread use for commercial oils. The oils are described as CA, CB, CC etc., the latest version being CG, and a supplementary number may be added, giving CG–4, for example. In theory all of these designations could still be used to describe oils for progressively more severe applications. In practice the categories CA, CB, and CC are no longer officially approved.

Roughly speaking, the later categories are suitable for more severe engine use, and CE, for example, would be suitable for all engines requiring oils from CA to CE. However, CD–II is intended particularly to cover severe two-stroke diesel applications.

The situation for spark-ignition ('Otto cycle' or petrol) engine oils is less complicated, since the military specifications for such oils have never been widely used. A similar API classification exists, progressing from SA to SJ with generally increasing severity, and categories SA to SF are no longer entirely current.

With the growing use of turbocharged, supercharged and petrol injection engines, the oil requirements came to be similar in some respects to those of diesel engines. From about 1986 certain oils have been labelled, for example, CD–SF, indicating that they meet both the CD and SF performance requirements.

Many of the larger vehicle manufacturers issue their own engine oil specifications, including Volvo, Daimler-Benz, General Motors, Ford, and Volkswagen. However, these are intended mainly to determine which oils are recommended for use in their various vehicles, and the oils themselves are usually quoted as meeting the API or MIL–L–2104 requirements. As a result the vehicle manufacturers' specifications are largely of interest to specialists, although their requirements can sometimes be critical for particular applications.

In an attempt to ensure that, in Europe at least, oils could be produced to satisfy the requirements of all the vehicle manufacturers, the Comité des Constructeurs d'Automobiles du Marché Commun (CCMC) also introduced a classification system, a recent specification level being D5. However, the CCMC has now become ACEA, the Association des Constructeurs Européens d'Automobile. ACEA introduced sequences TGO, TG1, and TG2 for petrol engines; TPD1 and TPD2 for diesel-engined cars; and TD1, TD2, TD3 and TD4 for heavy duty diesel engine lubricants. From the beginning of 1996, these were replaced by A1–96, A2–96, and A3–96 for petrol engines; B1–96, B2–96, and B3–96 for light diesels; and E1–96, E2–96 and E3–96 (and eventually E4–?) for heavy duty diesels. The situation is, therefore, extremely confusing, and is likely to be understood only by specialists. There is an urgent need for a clearer system which could again be understood by the non-specialist user. Table 11.13 shows the approximate relationships of some recent categories.

Table 11.13 Approximate relationships between diesel oil performance categories

	API	CCMC/ACEA	MIL–L–2104
High		D5	
	CF–4		
Performance		D4	2104 E/F
Level	CE		
	CD		2104 C/D
Low	CC		2104B

Finally, a recent development is the introduction in lubricant specifications of requirements controlling environmental effects. These include restrictions on certain additive types, for example to reduce heavy metal contents and chlorine, improvements in fuel economy, reduced oil consumption, and reductions in noxious emissions.

Apart from all these various performance specifications, engine oils are also classified in accordance with the SAE viscosity grades listed in Chapter 3.

11.6 Drawing up a specification

It sometimes happens that a lubricant specification is written to cover the requirements of a new application, before either the equipment or the relevant lubricants exist. In that situation it is also sometimes the case that no lubricant is available which meets all the requirements of the specification. It may then be necessary to compromise on the specification requirements or even to modify the design or operation of the equipment to enable available lubricants to be used satisfactorily.

As a rule, however, a specification is written to define the requirements of lubricants which are already known to exist and to be suitable for the application.

There are basically three ways of specifying a lubricant. The first is to define the exact range of composition, or formulation, which is acceptable. This is commonly done by the lubricant manufacturer, who writes a production specification for each product. A composition specification can rarely be used by the buyer of a lubricant, because the formulation will be kept confidential by the manufacturer. There are exceptions, where the composition is simple, such as in the British anti-seize specification DEF STAN 80–80 which contains equal parts of a well-defined graphite powder and of a soft petrolatum.

However, because of the importance of small variations in base oil composition, even the lubricant manufacturer will not normally rely purely on a composition specification, but will also specify quality control tests.

The second way to specify a lubricant is to require certain minimum results in functional tests, such as four-ball tests, gear tests, corrosion tests, and so on. Such tests are usually expensive to perform, and may be uneconomical for use with every batch of lubricant which is manufactured or bought. They may be used to check performance when a lubricant is first made, or when it is first submitted for an application, and may also be used for occasional random checks on subsequent batches.

The third way to specify a product is to place limits on a number of simple physical or chemical tests. These will be relatively cheap to carry out, so that they can economically be used for every production batch. They will not give any direct indication that the lubricant will perform satisfactorily, but they can ensure that a batch is similar to previous batches which are known to have performed satisfactorily.

A major specification for an important lubricant will often contain composition requirements, performance test requirements, and simple batch control tests. An example of such a specification is the British Defence Standard 91–30 for a graphite-containing lubricating oil.

The first thing to consider is the range of products which will be acceptable in the applications which are to be covered. If the lubricant is an oil, how tightly must the viscosity be limited? If it is a grease, what NLGI grade or what range of penetration is acceptable?

Once this is decided, there will probably be a clear indication of the properties of the lubricants which are suitable, and the specification can be designed around those lubricants.

(a) *Composition*

If all the base oils which are known to be suitable are mineral oils, and the user does not wish to become involved in the complications of assessing other types of base oil, then the base oil can be defined immediately as 'mineral oil' and this can perhaps be included in the title. Similarly if the choice is restricted to any other single type of base oil, that can be specified.

The best test to use to ensure that the right base oil is present is probably the infra-red spectrum.

On the other hand, it may not be considered desirable to restrict the type of base oil, in which case this composition aspect will not be stated.

Other composition factors can be defined as far as this is possible without unnecessarily limiting the choice of lubricant. For example a grease may be defined as 'mineral oil thickened with lithium soap, and containing anti-oxidants'.

It is not usual to specify the type or amount of additives, as these will have to be varied to compensate for variations in the base oil, but the amount of graphite or molybdenum disulphide may be defined as 3.0 ± 0.5 percent, or 3–5 percent, or 45–55 percent, depending on the type of lubricant.

(b) *Physical properties*

The acceptable viscosity range of an oil will usually be specified at a recognized reference temperature such as 40 °C or 50 °C. If the oil must be usable over a wide temperature range then either a minimum viscosity index will be required, or an additional viscosity limit will be defined at 0 °C or 100 °C. If the low temperature limit is critical, the pour point may be specified, for example as 'below −25 °C', or a viscosity limit at low temperatures in the Brookfield Viscometer or Cold-Cranking Simulator may be imposed.

For a grease, the penetration range or NLGI Grade Number will usually be stated. For the upper temperature limit, the dropping point is the best criterion. Remember, however, that some high temperature greases will not show a dropping point but will still be unusable above a certain temperature. If low temperature starting is important with a grease, a low temperature torque test can be specified, but the results of such tests are not precise.

Other physical properties will often be specified, such as specific gravity or relative density, colour, or even refractive index. These properties will probably not be important in themselves, but will be defined, and tested, to rule out the possibility of some completely different type of lubricant being introduced. Thus a change from one mineral oil to another, or to a synthetic oil, can often be detected by a change in physical properties.

Similarly, some of the simpler chemical tests, such as total acid number, total base number, or tests for a particular constituent such as calcium or zinc might be used to ensure uniformity of supplies or 'batch-to-batch uniformity'.

(c) *Functional tests*

For a critical application it is quite reasonable to require that 'any lubricant submitted as a candidate for this specification may be required to demonstrate satisfactory performance in the appropriate equipment in service'. In practice this service testing will probably only be required when a completely new type of oil is submitted. Such a test would constitute a 'type approval test'.

Other functional tests which are laborious or expensive would also be used only as type approval tests, including standardized engine tests. Some of the simpler and less expensive functional tests, such as four-ball extreme pressure, copper corrosion, dynamic anti-rust, or oxidation stability will also be used mainly as type approval tests, although they may occasionally be required on every batch where some special problem justifies it. A few functional tests, such as flash point and oil separation of grease may be carried out on every batch.

11.7 Precision of tests

Very few tests can be relied on to give exactly the same result every time.

If a skilled analyst or technician repeats a test with the same sample and the same test equipment, there will still be some variation between the results. The term 'repeatability' is used to describe the precision of the test under these conditions, and is defined as:

'that difference between two such single results as would be exceeded in the long run in only one case in twenty in the normal and correct operation of the test method.'

Where results are compared between different test operators in different laboratories, there will be a greater variation between results. The term 'reproducibility' is used to describe the precision under these circumstances, and is defined as:

'that difference between two such single and independent results as would be exceeded in the long run in only one case in twenty in the normal and correct operation of the test method.'

Most IP and ASTM tests quote repeatability and reproducibility figures which have been obtained by statistical analysis of lengthy correlation programmes. The repeatability and reproducibility can vary enormously from one test to another. Table 11.14 shows some simple examples.

Repeatability figures can be useful in that they enable an operator to check the precision of his own results against the published standard.

Reproducibility can lead to problems when a test method is not very precise. A lubricant supplier and a user may obtain results for a test such that the supplier's results show that the lubricant meets a specification while the user's results show that it fails. In most cases repeat tests will

Table 11.14   Some examples of test precision

Test method	Repeatability	Reproducibility
IP59 Method D	0.000016 g/ml	0.000016 g/ml
ASTM D 4049	6 percent	18 percent
IP50/ASTM D 217 (Worked)	7 units	20 units
ASTM D 3704	23 percent	39 percent

resolve the difference. However, for cases where a dispute cannot easily be settled, a recommended procedure can be found in British Standard BS4306 *Recommendations for the Application of Precision Data to Specifications for Petroleum Products.*

Chapter 12

Lubricant Monitoring

12.1 The objects of lubricant monitoring

Lubricating oils and hydraulic fluids are unique parts of a machine or system. They are vital components, whose condition is at least as important as that of other components. They are, however, unique because they can be sampled and inspected in great detail without damage to the machine and often without even interrupting its use. In operation the oil contacts other machine components, including many of those which are most critical. The changes in the oil can directly indicate the state of the bearings, gears, valves, and so on – these are called the oil-swept surfaces. Doing the right tests can therefore reveal a great deal about the condition of the plant, whether it is an engine, a gearbox, a rolling mill, or any other lubricated system.

It follows that monitoring of a lubricant can be valuable for two distinct reasons. The first is to provide information about the condition of the lubricant. The second is to obtain information about the condition of all the components which come into contact with the lubricant.

Systematic testing of lubricant samples from a piece of equipment is called lubricant monitoring. If it is used to keep track of the condition of the equipment it is a form of plant condition monitoring, or 'plant health monitoring'.

Some of the simplest methods of checking oil condition have been used for many years, but generally in a rather haphazard way. The systematic use of lubricant monitoring has largely developed since 1950, and it now involves some expensive techniques and complex organization. Such techniques and organization can only be justified where the lubricated equipment itself is large or expensive, or where safety problems are involved, but simple tests can be usefully applied to even small systems.

The removal of oil samples for analysis suffers from some disadvantages, such as the problem of finding a suitable sampling point, the labour involved in taking samples, delays in transporting samples to the place of analysis and sending the results back, and the obvious problem of obtaining representative samples.

As a result there has been a considerable demand for on-line lubricant monitoring techniques, and a wide variety is now available. In general, such on-line techniques are less informative than off-line techniques, but they have the major advantage of instant response.

The choice between on-line and off-line systems is not always simple. As in all things, safety and cost are major considerations. It may be sensible to install an expensive device giving limited information in a large passenger aircraft. It is, however, unlikely to be justified in a light aircraft or a road vehicle.

Similarly it may be appropriate to use even very sophisticated on-line equipment in monitoring a large power station, where only one critical lubricating system must be monitored. For a chemical plant or steel works, where a large number of separate and different systems must be monitored, it may be more cost-effective to bring samples to a central test facility.

There are two broad categories of test technique which are applied to samples removed from lubricating systems. The quality of the lubricant is monitored by a variety of chemical, physical, and sometimes even mechanical tests. The condition of the lubricated components is usually monitored by studying the nature and quantity of the wear debris carried in the lubricant.

12.2 Monitoring the quality of the lubricant

By taking a sample of the oil from a system from time to time and examining it, a great deal can be learned about the condition of the oil. This information can then be used as a basis for a number of decisions:

(a) if the condition is very bad, this may indicate that the oil should be drained and replaced immediately;

(b) if there are signs of breakdown, and the additives are becoming depleted, it may be possible to plan a shut-down for an oil change at some convenient time in the near future;

(c) if the oil and additives are not degraded, but there is solid contamination or water present, it may be desirable to drain it, clean it up ('launder' it) and return it to the system;

(d) if the oil is virtually as good as new, it may be possible to plan confidently to leave it unchanged for a very long time.

In other words, testing a sample of the oil may make it possible to use oil more economically, and as oil prices increase, this may become more and more important.

The processes by which a lubricant deteriorates are described in Chapters 3 and 5. The most important are oxidation of the base oil and additives, depletion of additives, and contamination. Many of the laboratory tests described in Chapter 11 can be used to monitor these processes. The choice of a suitable test or tests depends on the way in which the lubricant is likely to deteriorate.

12.2.1 *Over-heating and oxidation*

If a lubricant is overheated, the best indication is given by the presence of oxidation products. Many oxidation products of oils or additives, such as organic aldehydes or acids, contain a 'carbonyl' group, in which a carbon atom is linked by a double bond to an oxygen atom. This group produces a characteristic absorption band in the infra-red spectrum, as shown in Fig. 12.1, so that infra-red spectroscopy is a useful and inexpensive way to check for oxidation. However, it cannot be used in esters or other lubricants which contain a carbonyl group even when unused.

A more quantitative test for oxidation is to analyse for organic acids by the total acid number test in IP 139 or IP 177. At the same time the total base number test in the same methods can be used to show the amount of certain additives remaining (see Fig. 12.2).

More extensive oxidation can give tarry or solid oxidation products. These are insoluble in a light solvent called pentane, so that a test for 'pentane insolubles' can be useful in checking for severe oxidation. Many other solid contaminants are also insoluble in pentane and the results must be interpreted with care.

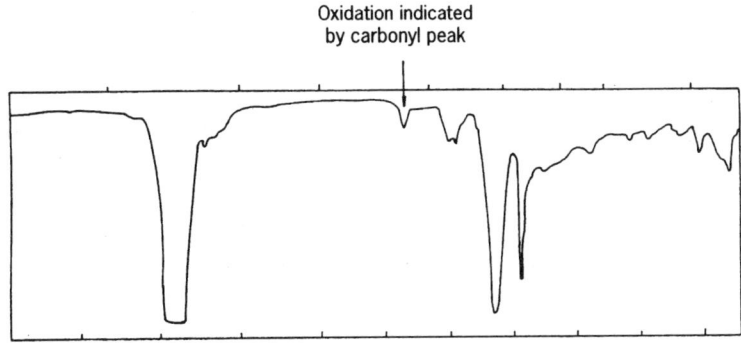

Oxidation indicated
by carbonyl peak

Figure 12.1   Infra-red spectrum showing oxidation

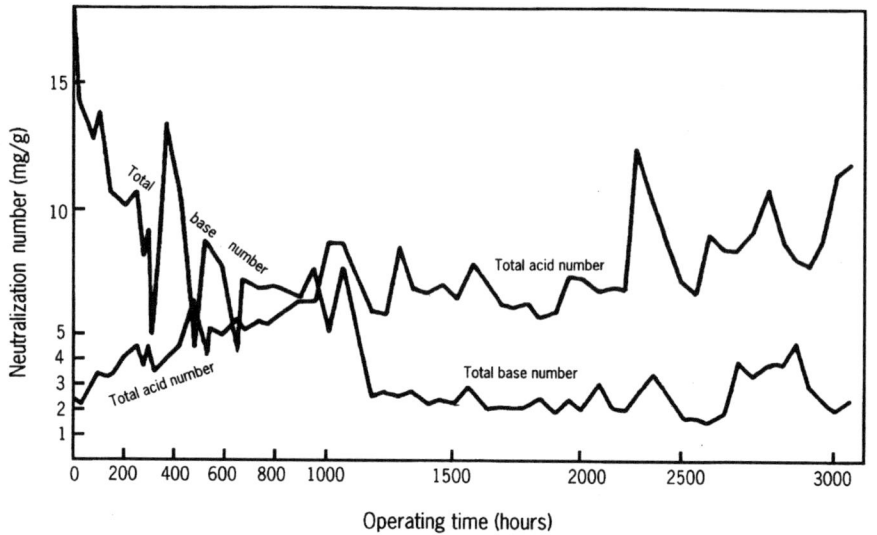

Figure 12.2   Neutralization number change in a diesel engine

If anti-oxidants are still present, and severe oxidation has not begun, it may be useful to assess the amount of anti-oxidant still present. This can be done by means of some of the oxidation tests mentioned in Section 11.3(a).

Oxidation will take place, to some degree, in any oil system. Tests must therefore be related to the time that the oil has been in use. If oxidation takes place unusually quickly it probably shows that some part of the system is too hot.

12.2.2   Depletion of additives

Many additives contain elements which are not found in the common base oils. Anti-oxidants usually contain nitrogen or phosphorus. Anti-wear and extreme-pressure additives contain sulphur, phosphorus, or – less commonly – chlorine. One very widely used class of additives is that of the metal dialkyl dithiophosphates, which contain sulphur, phosphorus, and a metal such as zinc, magnesium, or vanadium.

It can therefore be straightforward to monitor the depletion of such additives by analysing for the relevant elements. It must be borne in mind, however, that any of their breakdown products still present may also contain the same elements.

Institute of Petroleum and ASTM Standard Methods include tests for many elements, such as chlorine, nitrogen, phosphorus, sodium, sulphur, zinc, and other metals. These are based on colorimetric, gravimetric, polarimetric, potentiometric, and other techniques. All are accurate but more or less laborious, and therefore expensive.

Some additives containing metal elements, phosphorus, or silicon can be monitored accurately and cheaply by spectrometric oil analysis. This technique is mainly used for the monitoring of wear metals, and is therefore described in detail later.

It is also possible to test for depletion of additives by functional tests, such as those described in Chapter 11. This could theoretically be done for loss of anti-wear or extreme-pressure additives, using a test such as the four-ball test, or for a corrosion inhibitor by a static corrosion test; in practice, however, such a test would only be justified for a large volume of lubricant. A more practical example would be to use the foaming characteristics test IP 146 to test for depletion of an anti-foaming agent. Depletion of a viscosity-index improver can of course be assessed by the kinematic viscosity test IP71.

12.2.3 *Contamination*

A lubricant may become contaminated by almost anything, but the most important classes of contaminant are fuel (diesel fuel or gasoline), process fluids, solids, or water.

(a) *Contamination with fuel*
Fuel contamination can be caused by condensation of fuel in the cylinders of an engine at low temperature, or by an over-rich combustion mixture due to badly-adjusted carburettors or injectors, or simply by leakage, especially from diesel injectors.

There are some tests which are designed specifically to measure dilution of lubricating oil by gasoline, such as ASTM D–322 and IP23. Contamination by gasoline or diesel fuel can be checked by a fractional distillation test such as IP123/ASTM D–86 'Distillation of petroleum products', or by gas chromatography. All of these methods are relatively laborious and expensive, and are therefore not suitable for monitoring purposes. They can, however, be useful in failure investigations.

The most obvious effect of contamination of the lubricant by fuel is a reduction in viscosity. This can be monitored accurately and economically by a viscosity test such as IP 71. A typical record of viscosity variation for a diesel engine oil is in Fig. 12.3, which shows the effect of a fuel leak from an injector. For processing large numbers of samples, more recent automatic and semi-automatic methods are available with adequate accuracy and lower operating cost.

Flash point can also be used to check on fuel contamination, since even a small amount of fuel will produce a marked reduction in a closed flash point, such as IP 34. For more rapid testing of large numbers of

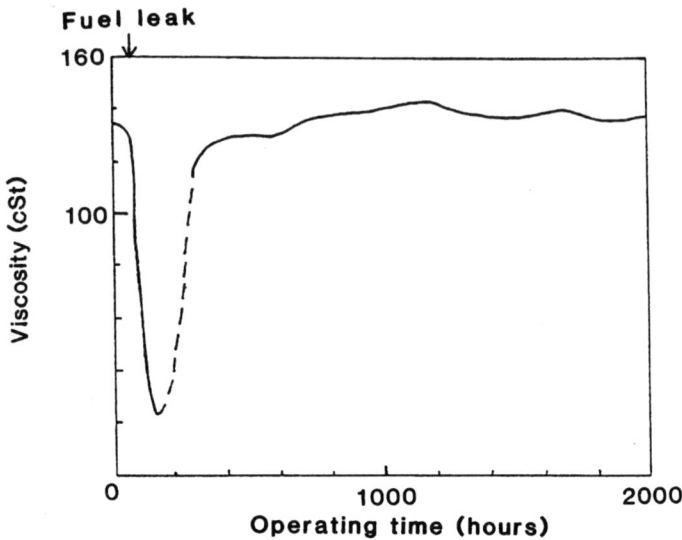

Figure 12.3   Viscosity monitoring showing effect of a fuel leak

samples an automatic flash point tester or a 'Go–No go' method such as IP 303 can be used.

(b)   *Contamination with solvents or process fluids*
Like fuels, most solvents or process fluids will reduce the viscosity of a lubricating oil and can be tested for by a viscosity measurement.

Similarly, a flash point procedure may indicate the presence of contaminants, provided that the contaminant has a flash point which is very different from that of the lubricant. Some contaminants such as chlorinated solvents or fluorinated refrigerants can completely suppress a flash.

(c)   *Contamination with solids*
The pentane insolubles test mentioned earlier is a useful test for most solid contaminants, because pentane is a good solvent for mineral oils and several synthetic oils but a poor solvent for most solids. Figure 12.4 shows the increase in pentane insolubles in a diesel engine in which poor injector timing was causing large amounts of sooty combustion products to be formed. This type of contamination can increase the viscosity of the oil and Fig. 12.5 shows the viscosity trace taken for the same engine.

An alternative method is to measure the benzene insolubles or toluene insolubles. Benzene and toluene are better solvents than pentane, and will dissolve most tarry oxidation products. The insolubles in these cases will, therefore, include a different range of substances from those insoluble in pentane. However, benzene is toxic; consequently, its use in

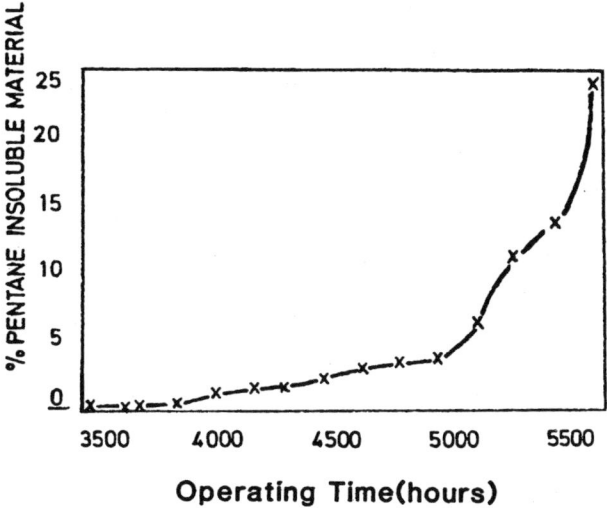

Figure 12.4   Increase in pentane insoluble materials in a faulty diesel engine

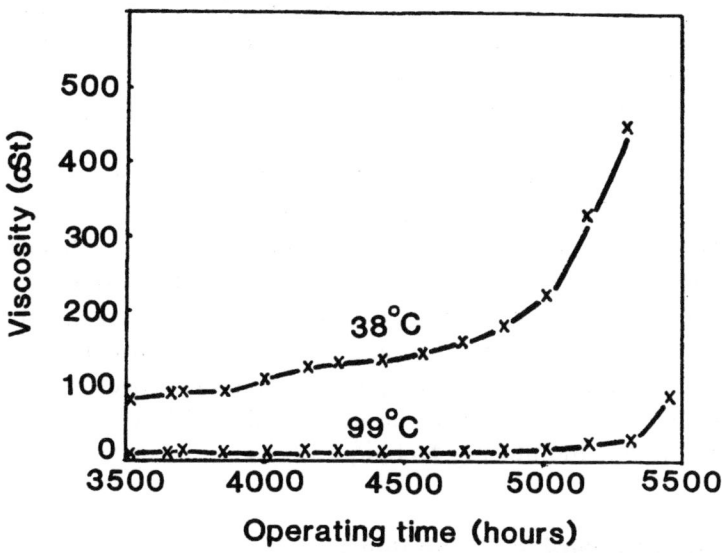

Figure 12.5   Viscosity increase due to sooty contaminants

laboratories is discouraged, and is subject to careful handling procedures.

(d) *Water contamination*
Lubricants can be contaminated with water from condensation, burned fuel, or leaks. One crude but effective test is to heat a sample of the

lubricant to above 110 °C. If much water is present the lubricant will make a bubbling or crackling noise like a deep fish fryer, and the test is commonly known as the 'crackle' test. (**Caution**: this test must be carried out carefully, not in a narrow or confined vessel, and with a gauze cover, and the heat should not be applied quickly. Water vaporizing rapidly can cause hot oil to escape from the vessel.)

A more precise test for water is called the Dean and Stark Method (IP 74/ASTM D–95), while for accurate measurement of very small amounts of water the Karl Fischer method is used.

In the special case of water containing a glycol anti-freeze, leaking into engine oil because of a damaged gasket or cracked block, the ASTM D–2982 test can be used to show the presence of the glycol.

12.3 Wear monitoring

The first systematic attempts to monitor wear could be said to have followed the introduction of magnetic drain plugs in oil systems. Magnetic plugs were introduced to remove ferromagnetic wear debris from oil systems and thus prevent further damage. Originally they replaced the ordinary drain plugs, and they could only be examined at the time of an oil change. It then became obvious from the debris present whether serious wear was taking place in the system. From this, the idea of adding a magnetic plug solely for wear monitoring was derived.

Further developments in magnetic plugs and filter systems, sometimes in combination, followed the simple plug, and these are described later.

The great development of lubricant monitoring since 1950 has been concerned with wear debris monitoring to monitor the condition of the lubricated equipment. The initial expansion was associated mainly with one technique, known as spectrographic or spectrometric oil analysis. This is probably still the most widely used technique. Spectrographic oil analysis is therefore described in the next section, followed by some of the other important monitoring techniques.

12.3.1 Spectographic oil analysis

In spectrographic (or spectrometric) oil analysis (SOA) it is the metallic wear debris which is analysed, and not the oil itself.

The technique was first investigated by some of the American railways in 1954. After some years of development it was found to give a very useful early indication of failures of diesel engines in railway locomotives. This enabled the locomotives to be withdrawn from service for overhaul in time to avoid many breakdowns in service.

In this way, the cost of repairs was very much reduced, and perhaps even more important, disruptions to rail services were also reduced. By 1965 the technique was regularly being used by British Rail.

In 1955 the United States started to examine the use of SOA for aircraft engines. It seems probable that it was the United States Navy which coined the expression SOAP, meaning 'spectrometric oil analysis programme', and this name is now in common use. Within a few years, while SOAP was still only being used on an experimental basis, it was considered by the United States Air Force that several catastrophic in-flight engine failures had been avoided by its use. At the same time, the overhaul life of jet and propjet engines was rapidly being extended. By 1968 the USAF had taken the remarkable decision to discontinue scheduled overhaul for some of its jet engines, and change to a system of on-condition overhaul. The engine condition was monitored mainly by SOAP. The decision to base engine changes on SOAP meant that the necessary analytical equipment became a normal part of the equipment of large USAF bases.

By 1970 SOAP was being used regularly by most large air forces for aircraft engine-condition monitoring, and by 1980 most airlines had followed suit. Other systems monitored included naval vessels and certain helicopter gearboxes. Due to a particular problem in one helicopter gearbox, those helicopters were not allowed to fly more than ten hours after the latest satisfactory SOAP analysis.

Other industries began using SOAP in the late 1960s, but adoption in industry has been slow and is still not widespread.

The two main instruments used to measure the concentrations of metallic elements in SOAP are the emission spectrograph and the Atomic Absorption spectrometer. Both depend on the fact that when atoms of an element are excited, such as by heating, their different energy states are separated by steps which produce particular frequencies or wavelengths and are characteristic of the element.

In the emission spectrograph a sample of the oil is ignited by an electric arc from a carbon electrode. The light emitted contains frequencies which are characteristic of the metals present, and the intensity is proportional to the concentration of each element.

In atomic absorption a sample of the oil is burnt in an acetylene flame. A concentrated beam of light is passed through the flame, and some of the light is absorbed at frequencies characteristic of the metal elements present. The amount of absorption at the characteristic frequency depends on the concentration of the element in the oil sample.

Thus in both cases a spectrum is obtained which indicates which metallic elements are present and in what concentrations. The results are provided as a table of metal concentrations in parts per million of oil. Some typical sets of results are shown in Table 12.1.

Table 12.1 Typical spectrometric oil analysis results

```
------------------------------------------------------------------
                EQUIPMENT   INTEGRITY   ANALYSIS
------------------------------------------------------------------
Swansea Laboratory Services,Department of Mechanical Engineering,University College of Swansea,Singleton,Swansea,SA2 8PP tel(0792) 295219

 File Name : paper.004.01                                 Date : 22 Oct 90
 Customer : An Integrated Steel Works.
 Operation :
    Unit No :                        Component : Gas Turbine
 Manufacturer :                      Model :
 Lubricant :
```

Sample Date		9 feb 89	19 mar 89	10 apr 89	2 may 89	19 may 89
Sample No		c1/-/7	c1/-/7	c1/-/7	c1/-/7	c1/-/7
Lubricant Hours/mileage		n/p	n/p	n/p	n/p	n/p
Equipment Hours/mileage		4433	4895	5255	5600	n/p
Makeup Lubricant		n/p	n/p	n/p	n/p	n/p
Wear	Iron	0	2	2	2	4
Elements	Chromium	0	0	0	0	0
	Aluminium	0	2	0	0	2
	Copper	4	5	11	15*	15*
	Lead	10	24*	59**	82***	114***
	Tin	0	2	2	0	1
	Nickel	0	1	0	0	0
	Manganese	0	0	0	0	0
	Titanium	0	0	0	0	0
	Silver	0	0	0	0	0
	Molybdenum	0	1	0	0	0
Additive	Zinc	99	48	42	33	55
Elements	Phosphorus	72	72	79	66	72
	Calcium	7	7	1	1	2
	Barium	23	3	1	1	1
	Magnesium	0	8	5	4	6
Contaminant	Silicon	0	2	0	0	3
Elements	Sodium	0	4	6	5	12
	Boron	0	0	0	0	1
	Vanadium	0	1	0	0	1
Physical	Fuel	–	–	–	–	–
Tests	Water	Neg.	Neg.	Neg.	Neg.	Neg.
	Dispersancy	–	–	–	–	–
	TBN/TAN	0.19	0.24	0.13	0.12	0.11
Ferrography	D1	2	24	21	8	8
	Ds	1	16	18	4	5
Wear Index	Is	3.0E0	3.2E2	1.2E2	4.8E1	3.9E1
Viscosity (cst at 40°C)		45	42	44	44	43
Result (Internal Ref No)		D1	D1	D19	D19	D19

```
Sample Date - 19 may 89
Condition:
        Critical - This Unit should receive immediate attention
Discrepancy:
        Abnormal wear trends suspected.
Recommendation:
        Inspect for main journal bearing wear as soon as possible.
        Monitor at reduced interval to maintain close surveillance.

* - Reportable , ** Unacceptable , *** - Urgent      Copyright - Spectron - Swansea Laboratory Services
```

A more modern emission technique uses a rotating disc electrode (rotrode). The preferred method now is probably an inductively coupled plasma (ICP) spectrometer, in which the sample is 'nebulized' or finely divided in a gas stream, and injected into a hot atomic argon plasma.

Modern instruments of either the rotrode or ICP types are typically microprocessor-controlled and computer-linked to give rapid sample handling and analysis, with on-line curve plotting and report printing. Some commercial laboratories provide direct data transmission to their clients, thus eliminating delays in supplying monitoring results. Detection limits for the various elements have also improved considerably. Table 12.2 lists some quoted detection limits for both types of instrument, although different analysts differ radically in their estimates of detection limits.

One problem with SOA is that large particles of wear debris will not remain uniformly dispersed in the oil. It can then follow that one sample will contain a few large particles, while the next sample from the same oil will contain none. The overall concentrations are usually small, and consequently the presence or absence of a few large particles can make a major difference to the result, and falsely suggest a major change in the wear situation.

To overcome this problem, it is common practice to filter or centrifuge the oil sample to remove the largest particles. This leaves the fine particles of metal or metal oxide, and the dissolved 'organometallics', reaction products between metals, and oil constituents. The results are then much more consistent, and the conclusions drawn are found to be

Table 12.2   Comparison of rotrode and ICP detection limits

Element	Rotrode detection limit (ppm)	ICP detection limit (undiluted sample) (ppm)
1. Iron	0.50	0.004
2. Silver	0.02	0.004
3. Aluminium	0.43	0.028
4. Chromium	0.72	0.007
5. Copper	0.09	0.003
6. Magnesium	0.01	0.0002
7. Sodium	0.14	0.015
8. Nickel	0.92	0.011
9. Lead	1.05	0.052
10. Silicon	0.38	0.018
11. Tin	1.80	0.055
12. Titanium	0.25	0.002
13. Molybdenum	0.79	0.005
14. Vanadium	0.62	0.003
15. Boron	0.09	0.0005
16. Barium	0.07	0.0005
17. Calcium	0.05	0.0002
18. Zinc	0.57	0.004

more reliable, although the individual concentrations are less accurate. It is also normal to compare the used oil with a sample of the same oil unused.

SOAP therefore gives figures which indicate the changes which have taken place in the metal concentrations. It does not give accurate absolute figures.

An individual SOAP analysis is of little value. The value of SOAP lies in continuous monitoring, which allows the operator to recognize changes in the condition of the system. With experience one can tell the importance of the changes, and decide when the equipment should be taken out of service. Figure 12.6 shows a graph of the change in concentration of some metallic elements over a long period in a diesel generator set.

SOAP will not distinguish the sources of an element, and metallic additives such as zinc compounds will show up in the same way as wear debris.

The advantages of SOAP are that is it cheap, needs only a small oil

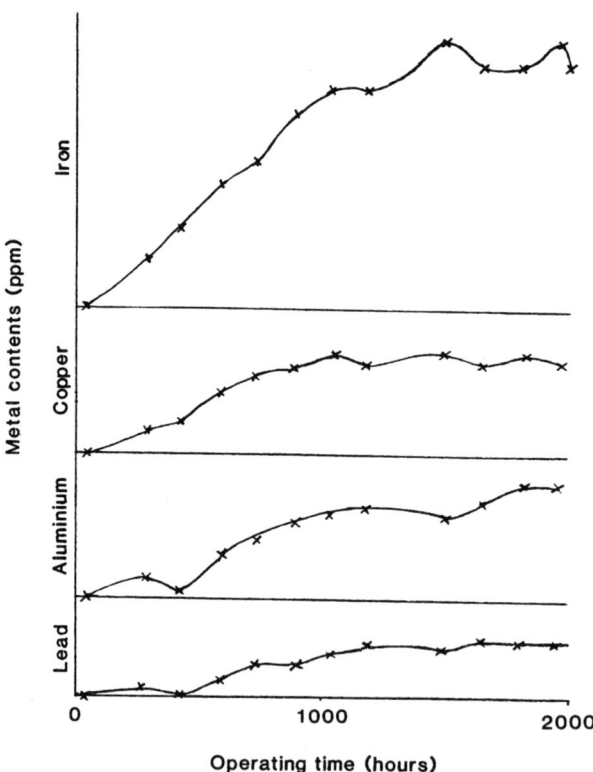

Figure 12.6   SOAP monitoring of diesel lubricant

sample, and can indicate to some extent the part of the system where something is going wrong.

The disadvantages are that it cannot distinguish between different types of damage (e.g., mild wear or corrosion) and may completely miss certain types, such as piston scuffing or rolling contact fatigue which produce mainly large particles.

Overall SOAP has been assessed by various users as between 90 percent and 98 percent reliable, depending on the type of equipment and the experience of the operators.

12.3.2 *Magnetic plug techniques*

The original magnetic plug consisted simply of a solid steel plug with a magnetized head, designed to screw into the sump of an oil system to replace the standard drain plug.

The plug was inserted into the sump, and it could, therefore, only be examined when removed in order to drain the system in the course of an oil change. It soon became obvious that the collection of wear particles on the plug gave a useful indication of the severity of wear taking place in the system, and hence that it would be useful to be able to examine the magnetic plug without needing to drain the system.

The next development was therefore to install the magnetic plug at some point in the system other than the sump. It was still necessary for the tip of the plug to be immersed in the oil. This meant that it was still not possible to remove it without loss of oil, and a further development was the design of a self-sealing fitting. This allowed the plug to be removed without losing oil, and this type is now in widespread use.

Magnets have also been used in conjunction with filters to improve the overall catching efficiency, but this complication seems not to have been widely used.

The first really successful on-line system for monitoring wear debris was produced by adding a pair of electrodes to a magnetic plug. Deposition of metallic particles on the magnetic plug eventually bridges the electrodes and activates an alarm. This device is known as a chip detector, and is in very wide use.

One problem with the basic chip detector is that the electrodes can be bridged by a few large metal flakes or by a large number of very small particles. The production of large flakes is always potentially dangerous. On the other hand, the formation of very small particles is only dangerous if the rate of formation is very high. To resolve the ambiguity of the chip detector, a small electric heater is added. When the electrodes are bridged and a warning is recognized by the operator, the heater can be switched on, and is powerful enough to burn off the fine debris, or 'fuzz'.

This 'burn-off chip detector' is therefore able to give the operator information about three different possible situations. If the warning is not cancelled by use of the 'fuzz burner' then the debris consists of large chips or flakes, and is potentially dangerous. If the warning is cancelled by the 'fuzz burner' but is rapidly re-activated, then the debris is fine but is being generated rapidly; this again is potentially serious. Finally, if the warning is cancelled by the burner and does not recur, then the debris is fine and the rate of production is slow, so that there is not likely to be any immediate cause for concern. This is probably the most widely used on-line wear monitoring system at present.

A modification of the standard burn-off chip detector is the Zapper*, in which a burn-off pulse is automatically emitted when the electrodes are bridged. This pulse is sufficient to break the electrical continuity if fine debris is present, but leaves the bulk of the debris in position for later examination. By breaking the electrical continuity the Zapper prevents nuisance warnings caused by harmless fine debris. Such 'nuisance' warnings are no more than an irritation to operators of industrial plant, but may cause dangerous distraction to an aircraft pilot.

If the Zapper technique is arranged to prevent any warning indication to the operator from fine debris, then it may mask the very rapid generation of fine debris which can sometimes occur and which is itself potentially harmful. It may, therefore, be better to arrange that the Zapper allows a brief transient warning. A single indication could then be ignored, but repeated activation, such as with continual flashing of a light, would be a matter for concern and investigation.

The most sophisticated developments of the magnetic plug are the Tedeco quantitative debris monitoring (QDM) system* and oil debris detection system.

*Zapper and QDM are registered trademarks of Tedeco.

12.3.3 *Ferrography*

The Ferrograph* is a laboratory-based instrument which in its commonest form provides two distinct wear-monitoring techniques – the Analytical Ferrograph and the Direct-Read Ferrograph. Both depend on the use of a strong magnetic field to deposit wear debris from a sample of used lubricating oil. Figure 12.7 shows the standard Duplex ferrograph, comprising both techniques.

In the Analytical Ferrograph an oil sample is pumped by means of a peristaltic pump to flow onto a substrate mounted on top of a powerful permanent magnet. The substrate is a standard microscope slide on which a barrier film has been deposited to constrain the oil to flow axially along the substrate. The majority of the ferromagnetic debris

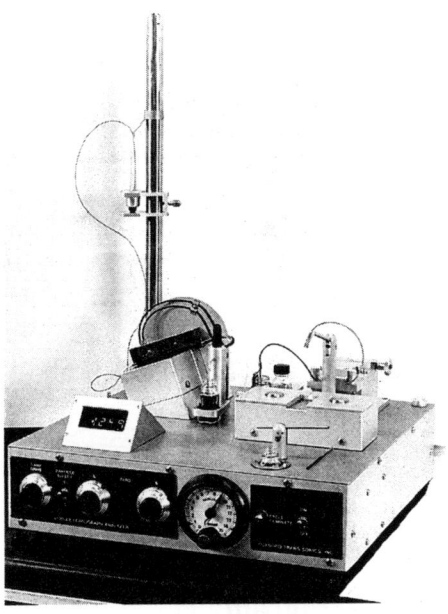

Figure 12.7   Duplex ferrograph

present, and probably some other magnetically susceptible debris, is influenced by the magnetic field, which causes it to be deposited on the substrate.

The substrate is then washed to remove all the oil. The debris remains on the surface, where it is held quite strongly. The deposit on the substrate, known as a ferrogram, is in a very convenient form for microscopic examination, either by optical or electron microscopy.

Ferrography was the first simple, convenient technique which enabled wear debris to be examined to establish the type of wear taking place and the severity of specific wear mechanisms.

In theory the more strongly susceptible ferromagnetic particles are deposited first, thus giving a separation of different particle types along the axis of the substrate. In practice this separation is not very efficient. A typical Ferrogram is shown in Fig. 12.8.

In the Direct-Read Ferrograph a similar procedure is used to produce a deposit of wear particles in a glass tube. Optical density measurements are then made at two points on the tube representing larger and smaller particles. The two values thus obtained can be plotted independently for trend monitoring purposes, or can be used to calculate a parameter called the severity of wear index which can be used in the same way.

*Ferrograph is a registered trademark of the Foxboro Corporation.

Figure 12.8   Typical ferrogram

12.3.4   *The rotary particle depositor (RPD)*

The Analytical Ferrograph is a convenient technique for separating wear debris from an oil sample for analysis. Experience showed, however, that it had some deficiencies.

The RPD was designed to overcome the problem of particle overcrowding at the entry point. The substrate, which is a microscope cover-slip, is mounted on a rotating magnet assembly consisting of an inner solid cylindrical magnet and an outer hollow cylindrical magnet with reversed polarity. The oil sample is made to flow onto the substrate at the centre of rotation, so that it flows radially outwards over the whole surface of the substrate. The positions of highest magnetic gradient are at the circumference of the inner magnet and the inner and outer circumferences of the outer magnet.

The debris is therefore deposited in three concentric rings at those locations. By appropriate selection of the speed of rotation and of the field strengths, the particles are made to segregate themselves in the three rings in accordance with their magnetic susceptibilities. The largest and most highly ferromagnetic particles deposit in the innermost ring, while the other rings contain particles of progressively decreasing susceptibility. At the edges of the magnets the gradients are very high, so that particles of very low magnetic susceptibility are deposited, and the overall capture efficiency is high.

A rotary particle depositor is shown in Fig. 12.9, and some typical deposits of wear debris in Fig. 12.10.

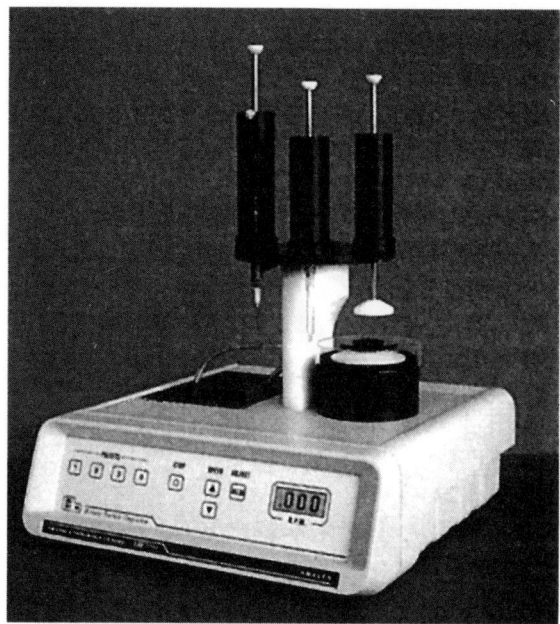

Figure 12.9   Rotary particle depositor

12.3.5 *Other monitoring techniques*

Many other wear monitoring techniques are available which have advantages and disadvantages for particular applications.

The Inspection Instruments Debris Tester uses an eddy current technique to obtain a quantitative measure of metallic debris. The debris must first be concentrated on a non-metallic substrate. The normal technique is to filter the required quantity of oil through a fine filter and then measure the quantity of debris on the filter. An alternative technique for engines and gearboxes is to remove the accumulated debris from a conventional magnetic plug by means of adhesive tape, and then attach the tape to a card from which a reading is taken. The card can then be stored to provide a monitoring record.

Both procedures are cumbersome and labour-intensive. Nevertheless, the technique has been widely used, especially in the mining industry and in military aircraft.

The Particle Quantifier (Fig. 12.11) is one of the few techniques which can measure the quantity of debris present in an oil sample without any preliminary separation. Alternatively, the debris can be in the form of an RPD deposit, on a filter, or separated from a specially designed magnetic plug. It uses a high-frequency AC magnetometer. The reading obtained is strongly influenced by the magnetic susceptibility of the debris, but the results are suitable for trend monitoring.

A direct indication of the abrasiveness of solid particles present in an

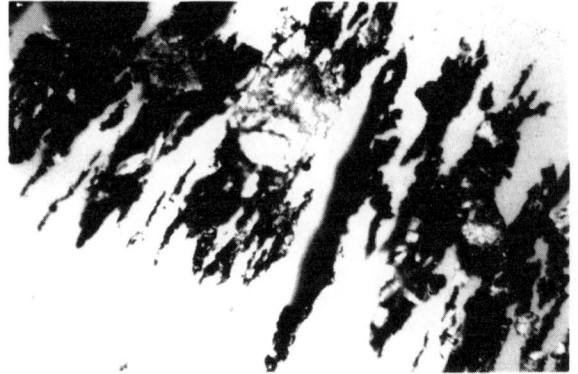

Inner Ring of slide × 128

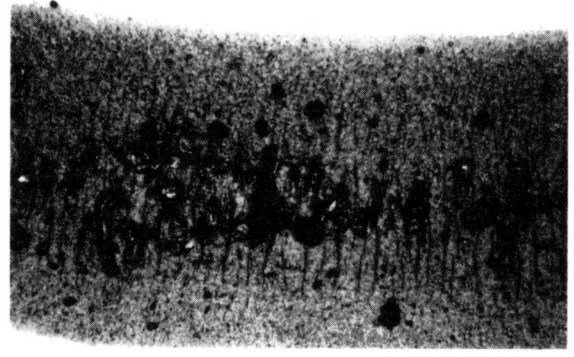

Outer Ring of slide × 128

Figure 12.10   Typical RPD wear debris deposit

Figure 12.11   Particle Quantifier ferrous debris motor

oil can be obtained with an instrument from the Fulmer Research Institute. This must be considered more as an oil quality assessment than as a wear monitoring device, since the abrasive may be from airborne dust or any other source.

12.4   Field tests

The laboratory-based tests described in the previous sections are of great value for large or critical systems, and where changes are relatively slow. The on-line techniques are valid only for critical systems. In many situations simpler tests are needed which are cheaper and can be carried out on the spot. Such tests are naturally much less precise, but can be very useful for small systems such as goods vehicles or marine diesel engines.

(a)   *Blotting-paper test*
If a drop of oil is placed on a sheet of blotting-paper or thick filter-paper, it will spread slowly, giving a more or less circular spot, or blot. A new oil without any insoluble additives will give a uniform pale yellow, slightly transparent spot. An oil with solid additives or contaminants will leave them as opaque flecks near the centre of the spot. A degraded oil, or one containing dark contaminants, will give a darker brown spot, or may produce one or more brown or black rings. Some typical blotter test spots are shown in Fig. 12.12.

(b)   *Viscosity tests*
A simple viscosity test can be done by any technique which allows a sample of the oil to flow. The enormous effect of temperature on viscosity means that it is easy to be misled about a change in viscosity of the oil. Whatever test is used should be done in comparison with a sample of the same oil in its new condition.
    One fairly sophisticated example of the on-the-spot viscosity comparison is the Mobil Flo-Stick, shown in Fig. 12.13. This is a plastic device consisting of duplicated reservoirs and flow channels. One reservoir is filled with the used oil and the second with the equivalent unused oil. When the Flo-Stick is tilted, the two oil samples run down their respective flow channels. When the unused sample reaches a mark on its channel, the used oil is considered to be in a satisfactory condition if it has reached a point somewhere between two marks on its channel. Within the accuracy of the device, this is equivalent to saying that the viscosity of the used oil is satisfactory if it is within plus or minus x % of that of the unused oil.
    Even simpler techniques simply involve watching the used and unused samples flow through a small funnel, or drip from a glass tube, and estimating how much difference there is between them. These simpler

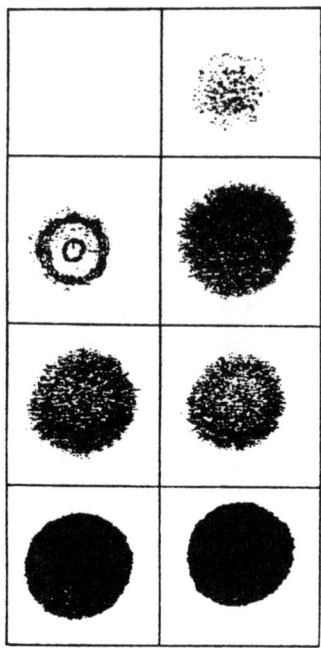

Figure 12.12   Blotter tests

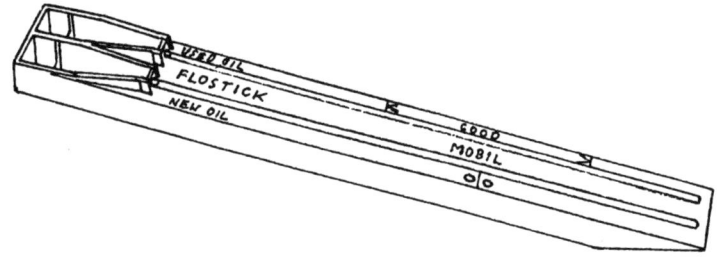

Figure 12.13   Mobil flostick

techniques are, of course, very imprecise, especially if solid contaminants in the used oil disturb its flow, but they do show up major thickening or fuel contamination.

(c) *Capacitance tests*
There are several devices, like the Hoppe An-Oil-Iser, in which a static electric charge is generated between two plates immersed in the used oil, and the capacitance is measured. The presence of water, metallic debris, or acidic oxidation products will reduce the capacitance.

(d)   *Appearance*
With careful observation, a lot can be learned from the visual appearance of the oil, preferably in a small clean glass container, but if necessary even on the surface of a dipstick. Darkening, thickening, carbonaceous deposits, and water may all be detected by eye, and with practice the operator can make a sensible decision about the state of the oil.

It is always worthwhile to look at the oil which has been drained from a system. Wear debris, combustion products, water, and sludge can be easily seen and will give a clear indication that a more detached look should be taken, either at the oil itself or at the lubricated system.

12.5   Application to grease

There are several problems in applying lubricant condition monitoring to grease-lubricated systems. The first is that it is more difficult, and sometimes impossible, to obtain a sample from an operating system. It can be done where there is a grease nipple or centralized feed, and where surplus grease exudes from a grease valve or seal.

The second problem is that because there is virtually no flow in a grease system, there will be enormous variations between samples from different places. For example, wear debris or oxidation products may be present in only one small part of the system, and monitoring which is based on samples from elsewhere will completely miss them. It is therefore important that a sample for analysis should be representative of the grease at the working surfaces.

Compared with those two problems, the difficulties in analysing greases are minor. A grease can always be dispersed in clean base oil or solvent, and can then be treated in the same way as an oil. There are, of course, particular tests, such as penetration or dropping point, which can be applied directly to a grease sample. Changes in these properties can give useful guidance as to whether the system should be re-greased more or less frequently.

In practice, grease monitoring is used far less than oil monitoring. One reason for this is that the large expensive or critical systems are more likely to be oil-lubricated.

Analysis of used grease may be useful in investigating bearing failures. This is considered in Section 12.7.

12.6   Selecting and designing a monitoring programme

Lubricant condition monitoring can be expensive: equipment failures can be even more expensive. The objective in designing a monitoring programme is to obtain the best possible value for money, in other words to get the best possible prediction of problems for the least possible monitoring effort and expenditure.

The probable cost of a failure will include:

– direct cost of repairs or replacement parts;
– cost of maintenance effort;
– cost of down time (consequential losses);
– risk to life, limb, or property.

The last item cannot be costed in simple cash terms, but may be the most important part of the overall risk.

The second factor to consider is the likelihood of failures occurring. This involves:

– frequency of failures (mean time between failures);
– nature of failures, and frequency of each type;
– measures already taken to prevent failures.

This last item can be a complicated one, including such measures as frequent lubricant changes and inspections or overhauls, operating at reduced capacity, existing monitoring devices, and so on.

These first two factors – cost and risk of failure – together provide the justification for a condition monitoring programme, but neither the cost nor the risk is easy to assess.

It may be considered, for example, that the cost of failure of a big-end bearing in a heavy goods vehicle would be £1000 including repair and lost revenue, and that the frequency of such a failure is one in a hundred in one year. This would indicate that even if lubricant monitoring could enable all such failures to be avoided, it would not be worth more than £10 per year to monitor every vehicle.

On the other hand, the cost of a brake failure, or even of a seized piston or main bearing, may be enormous. The engine of a truck in North America once failed while it was on a level crossing, and an express train was de-railed at a cost of over $10 million. Similarly, in large petrochemical works, failures of bearings or hydraulic systems have been known to cause catastrophic fires costing many millions of pounds.

Such incidents are so rare that they are difficult to include in the equation. In this respect condition monitoring is like insurance, in that a small regular expenditure may be a way to avoid occasional catastrophes. In fact it is perhaps surprising that the major industrial insurance companies have not been more in the forefront of promotion of plant health monitoring.

Having, by calculation or guesswork, decided how much it is reasonable to spend on monitoring, it is then necessary to assess the method of monitoring which can be used, the probability of that method preventing the failure, and the frequency of testing – all within the expenditure which can be justified.

It may be interesting to illustrate this problem by means of examples, bearing in mind that this book is concerned with monitoring based on the lubricant, and not on other monitoring techniques, such as temperature and vibration monitoring.

A classic case is that of a large jet-propelled airliner, carrying 350 people on the North Atlantic route. It has four engines, and can fly quite safely on three, or even, within certain limitations, on two. The risk of an engine failure is therefore not a risk to life, but a risk that the aircraft may have to be diverted, so that the cost of the flight will be wasted. The cost of the flight may be £50 000, and the mean time between failures for the engines 10 000 hours, or 1500 North Atlantic crossings. If the probability of SOAP analysis preventing an engine failure is 80 percent, it is well worth spending £5 to have a SOAP analysis carried out on the oil from each of the four engines before every flight. On the other hand, it would not be worth carrying out analytical ferrography at £20 each on every engine before every flight.

At the other end of the scale, a large articulated lorry is shipping goods from Britain to Italy. The overall cost of an engine failure en route is estimated as £1500, and the risk of engine failure from all causes is 1 in every 100 return journeys. If a detailed lubricant analysis costing £40 stood a 60 percent chance of preventing en-route failures, it would not be worth doing. If a simpler analysis costing £6 stood a 40 percent chance of preventing failure, it might just be worth doing. If an examination of the dip-stick and a blotting-paper test, costing £2 in mechanic's time, stood a 20 percent chance of preventing failure, it would be worth doing.

Finally, consider the problem of a pump bearing in a petrochemical plant, where the pump is handling an inflammable solvent. Suppose there are 5000 such bearings in similar duty in Britain and that 200 of them fail every year. The normal average cost of a failure is £800, but once in five years such a failure starts a fire which costs £500 000. The annual cost per bearing is then

$$\frac{(800 \times 200) + \dfrac{(500\,000)}{5}}{5000} = £52$$

It will, therefore, be worth spending up to £52 per bearing per year for a monitoring technique which will completely prevent failures.

One other aspect affects the frequency of monitoring. This depends on the time it takes for a particular failure mechanism to develop from the detectable stage to failure. Wear of a white metal bearing might take 300 hours to progress from the first detection of tin in the oil to complete failure; it will therefore be useful to carry out a SOAP analysis for tin once per week. A roller bearing may suffer rolling contact fatigue, and progress from the first wear debris to seizure in 40 hours. It would be necessary to carry out analytical ferrography every day to avoid seizures, and a better solution would probably be a vibration detector.

A less evident problem in designing a monitoring programme is to define what indicates failure. In some cases this may be obvious, for example the sudden appearance of severe wear particles in an oil. In many cases, however, there is a steady increase or decrease in the property being monitored, as in Figs 12.4 and 12.5, and there is no obvious sudden change.

Reliable correlation between a monitoring level and the probability of failure can often only be established by experience, and in many cases the decision that a failure is imminent is very subjective.

It is common practice to establish two limits. The first is called a reporting, warning, or surveillance limit, and when the monitored property reaches this level, more intensive monitoring or observation of the equipment is carried out. The second limit is the removal or rejection limit, and when that level is reached the equipment is normally taken out of service.

12.7   Investigating failures

A failure takes place when the material of a piece of equipment is subjected to stresses which it cannot withstand. The stresses may not just be excessive mechanical stresses – they may include overheating, corrosion or other types of chemical attack. It follows that if enough is known about the properties of the material and the stresses it has experienced, it will be clear why the component failed.

The general subject of plant failure analysis is outside the scope of this book, but some of the techniques of lubricant monitoring can also be used in failure analysis. One problem with bearing failures is that often the bearing will have been virtually destroyed, so that all the evidence in the bearing surfaces of the nature of the failure will have disappeared. When that happens, the lubricant may still contain particles generated at all stages of the bearing's service. Analytical ferrography of a sample of the lubricant may then show the whole history of the type or types of wear which took place leading up to the failure.

Some other examples of the use of lubricant tests for failure analysis are:

(a) a penetration test may show that a grease had hardened so much that it could no longer supply the bearing surfaces;

(b) infra-red examination of an oil or a grease may show the presence of contaminants, or oxidation products from overheating;

(c) filtration of an oil, and examination of the residue with a microscope may show contamination by an abrasive, such as grinding or grit-blasting grit, or sand;

(d) a viscosity measurement may show that the oil is contaminated with fuel or a solvent, or that the wrong oil was used.

Failure investigation requires an open-minded approach, and the use of all techniques which may give useful information. Each investigation is different, but careful thought will usually indicate which tests are likely to be helpful in a particular case.

The most important thing in any failure investigation is not to disturb or discard any component until it is clear that by doing so vital evidence will not be destroyed. This applies particularly to the lubricant.

Chapter 13

Lubricant Handling and Storage

13.1 Care in lubricant handling

Bearings are almost always precision components, and the bearings, seals, valves, and other lubricated items are the most critical in many machines. The lubricant is a critical part of such components, and should be handled with as much care as any other piece of precision engineering.

Just as any metallic item can be degraded by heat, moisture, chemical attack, or the presence of inclusions, so can a lubricant be damaged by heat, cold, moisture, chemical attack, or solid contaminants. Deterioration of a lubricant will be passed on to bearings, gears or any other lubricated parts.

It follows that it is just as important to handle and store lubricants carefully as it is for any other part of a machine. This is not always sufficiently understood, and although the standard has improved enormously during the past twenty years, there is still a tendency in many organizations to treat lubricants with far less care than other items.

Lubricant manufacturers have made great efforts to handle the problem by ensuring that when lubricants are sold they are as clean as possible and are packaged to withstand mechanical or physical damage. However, all their efforts are wasted if their products deteriorate due to careless handling or storage before they reach the machines.

13.2 Lubricant packaging and delivery

Lubricants are supplied in many different types of package, and the most suitable type depends on the quantity used and the application. The amount to be purchased at any one time is discussed in Section 13.3 in connection with storage.

Oils can be supplied in quantities varying from tiny capsules or bottles of a few grams to bulk deliveries of several thousand gallons. Bulk deliveries are usually made by road or rail tanker, the minimum volume being perhaps 500 gallons for a small road tanker. Delivery may be made direct to the reservoir of a large centralized lubrication system or to a storage tank. It is the responsibility of the supplier to make sure that the tanker and its pumps, hoses, and couplings are clean. It is the user who is responsible for making sure that their tanks or reservoirs, and hoses and couplings are clean and serviceable.

If the oil is identical with a previous batch in the same system, then cleanliness will be the only concern, since any unused oil remaining in the system will not degrade the new oil. If the new oil differs significantly from the previous oil, it may be important to drain and flush the system before the new oil is introduced.

It is worth noting that 2.5 m of 50 mm (8 ft of 2") bore hose will hold 5 litres (over a gallon) of oil, while the film remaining on the walls and bottom of an 'empty' 500 gallon tank may contain several gallons of a heavy oil. When the type of oil in a storage tank is changed, the tank and its associated equipment should therefore be thoroughly drained, and either cleaned or flushed throughout with a small quantity of the new oil and again drained. Cleaning may be by steam-cleaning or solvent cleaning. In either case, all the equipment must be completely ventilated and dried before the new oil is introduced.

Some manufacturers will supply up to 500 gallons of oil in a portable tank. This is a covered rectangular tank designed for easy handling by a hoist or a forklift truck, and fitted with a sight glass and a tap. It is delivered full and installed on a raised platform in the lubricant store or 'cellar', or in a suitable position in the factory. It is then used as a storage tank, from which oil is supplied to dispensers or oil-cans. When empty, it is replaced by a full tank and returned to the supplier for re-filling.

The most common package for large quantities of oil is the standard European size steel drum or barrel containing 209 litres, equivalent to the old 45 gallon size. These drums are fitted with two screw plugs in the flat top, and the caps are carefully sealed when the drum is filled. The larger cap can be replaced by a tap or pump so that the drum can be used as a storage and dispensing unit.

Smaller drums and cans containing 1, 2, 5, 10, or 25 litres are used to supply smaller quantities of oil for industrial use or for retail supplies of oil to domestic users.

The smallest containers are used for the sale of specialty products such as watch, clock, or instrument oils, and branded retail products for such applications as sewing-machines, bicycles, fishing reels and rod ferrules, shot-guns, and skateboards. The containers may be bottles or small disposable oil-cans made of glass, metal, or plastic. The price per gallon of the oil sold in these small vessels is far higher than in larger containers. Oil should never be bought in this way unless only a very small quantity is needed.

Greases are normally sold for industrial use in kegs or barrels containing 3–180 kgs (7–400 lbs); smaller cans containing 0.5 kg (roughly 1 lb) are available if the usage is small. Some greases are sold in cartridges which are designed to fill a standard size grease-gun. As with oils, specialty grease products are supplied in very small packages containing less than 100 gms, and these are usually tubes made of thick metal foil or polypropylene.

The polypropylene tubes are less likely to rupture than the metal ones, but do not remain collapsed after squeezing, so that they may draw air and dust in when they recover their shape. If they are used over a long period, the grease content can become oxidized, hardened, or dirty.

All types of lubricant container must be handled carefully to avoid damage. This applies particularly to the large oil drums and grease barrels which are difficult to handle carefully because of their size and weight. Drums can be rolled safely over flat surfaces, but should never be allowed to drop more than an inch or two because the seams may become strained and leak. Over rough ground or steps they can be rolled smoothly on two lengths of timber set about 2 feet apart. For unloading from lorries, proper hoists, fork-lift trucks, or barrel skids should be used.

13.3 Storage

The first consideration in lubricant storage is to avoid storing too many types of lubricant. In other words, lubricant rationalization, as described in Chapter 4, should be carried out before the storage system is designed. If possible, the store should then have some spare capacity to cope with additional needs which may arise later.

The second consideration is how much of each type needs to be stored, and in what size containers. It is difficult to give any hard and fast rules for this. Oil in a clean sealed drum, kept in a dry place, may remain in perfect condition for many years, while most greases are stable for at least two years. Once the containers are opened, however, deterioration is more rapid. The size of container should probably be chosen to represent between one and three months supply.

Ideally the stores should have room for at least two containers of each type of oil or grease, unless large storage tanks of 600 litres (roughly 150 gallons) or more are used.

Some of the factors involved in selecting the number and size of containers are:

(a) it is generally more economical to purchase and to store in large quantities;
(b) small containers are emptied more quickly, and are therefore less susceptible to contamination;
(c) small containers are easier to handle.

Once the total quantity of lubricant to be stored has been decided, it is possible to plan the location and layout of the stores area.

Lubricant stores should always be under cover. Typical locations include a corner of the enclosed factory area, part of the general stores area, a lubricant 'basement' or 'cellar', or a separate building away from the main factory building.

Where the overall quantities of lubricant are small, location in the factory building or in the main stores is usually preferable.

Where quantities are large, and continuous supplies are made to integrated machinery such as rolling mills, cellar storage has several advantages. A cellar can be centrally located near to the use areas, but the lubricant storage is kept out of the way of production equipment and workers. A minor advantage is that storage tanks can be filled by gravity from road or rail tankers at ground level.

Lubricant storage is often combined with the storage of solvents and paints. The overall fire risk is then quite high, and storage should preferably be in a separate building with adequate fire precautions. Such an arrangement will probably result in reduced insurance premiums for the main factory buildings.

Smoking and naked flames should always be banned in any lubricant storage area, and the bans should be enforced. Access to lubricant stores should also be restricted to a very few people authorized to issue and dispense the lubricants. Troubles will always arise if machine operators are allowed to draw their own lubricants, unless they have been given clear guidance on the need to use the correct specified product, and the importance of cleanliness.

The lubricant store should also have floor gratings and drainage channels to allow any spilled oil to drain away and reduce floor contamination.

Large oil storage tanks are usually rectangular, as this allows more economical use of space. They may be mounted a few inches above the floor, just enough to prevent moisture from corroding the tank bottoms, or they may be mounted up to three feet above floor level. The latter permits small dispensing vessels to be filled from non-drip taps by gravity, whereas the oil has to be pumped out of a floor-level tank. In either case, the tank should be tilted slightly towards the front to enable it to be emptied completely. A typical installation of large storage tanks is shown in Fig. 13.1.

Large storage tanks may be filled directly from a tanker or from barrels, either by pouring or by pumping.

For storing intermediate quantities of oil, drums are very convenient containers. The best arrangement is probably to mount them horizontally in racks, fitted with non-drip taps to enable dispensing vessels to be filled by gravity. A typical rack system is shown in Fig. 13.2. Individual drums can also be mounted on barrel stillages or on barrel trucks, which allow them to be stored upright and tilted horizontally for dispensing.

Drums can be stored vertically, and equipped with a hand-pump to dispense the oil. They should preferably be raised slightly off the floor to prevent moisture from accumulating underneath and corroding their bases. If outside storage of drums cannot be avoided, or if there is any risk of water falling on them, they should never by stored upright.

Water can collect in the recess in the top, and even with first-class

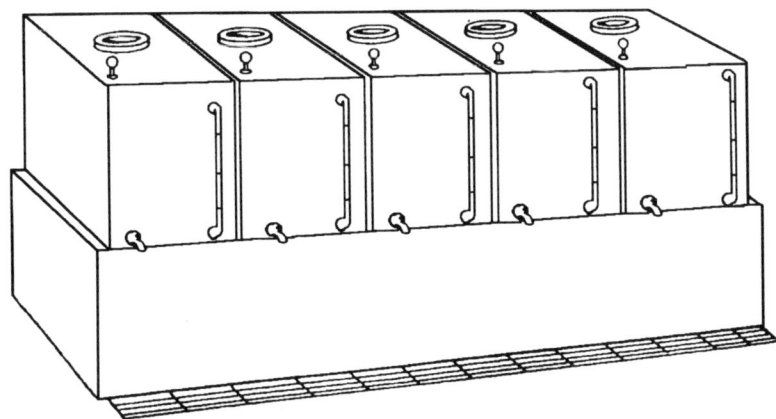

Figure 13.1   Bulk oil storage tanks

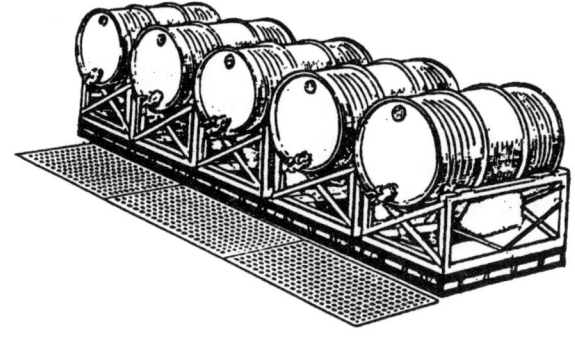

Figure 13.2   Barrel rack storage

sealing it will eventually enter the drum. As the drum warms up during the day, the contents will expand, and a small quantity of air will be forced out through the plugs. When the drum cools again at night the contents contract, and this creates a partial vacuum which will suck in water surrounding the plug. The water then sinks to the bottom of the drum, and the process can be repeated indefinitely. Drums which have been stored for years out of doors have sometimes been found completely full of water, with the seals still intact.

It is generally unavoidable for grease barrels and kegs to be stored upright, and again wherever possible they should be stored under cover. The lids of grease containers are less deeply recessed than those of oil-drums. Even so, if outside storage is unavoidable, the kegs should be tilted slightly to enable water to drain off, and also covered with a tarpaulin or plastic sheet.

Small tins, cans, bottles, and tubes of lubricant are most conveniently stored on racks or shelves, or in cupboards – in the same way as other precision parts.

The careful and clear labelling of lubricant containers is even more important than for other components. If the label on a mechanical component becomes unreadable, or missing completely, its size and shape will often identify it, or at least prevent it from being used in the wrong place. A lubricant has no such unique features to identify it, and a lubricant which has been wrongly identified may be used in the wrong system with catastrophic results. A lubricant container which is unlabelled should be discarded.

Equally important is date identification of lubricant containers. Sealed drums or cans of oil should be scheduled for use within a maximum of one year of purchase, and greases within six months. For specialty materials, such as some of the synthetic oils and calcium soap greases, the period should sometimes be even shorter. They will, of course, in most cases keep for much longer periods, but deterioration will continue at a faster rate once the container is opened.

The main object of date labelling is not to fix a maximum storage period, but to ensure that containers are used in the same sequence as they are bought. A container which sits out of sight at the back of the store for several years before being used may cause more trouble than it is worth, and should also be discarded, or at least used only for non-critical applications.

In addition to lubricants the store should also accommodate all the dispensing equipment – pumps, charts, and even the cleaning rags which will be used by the lubrication staff. The cleanliness of all these items is just as important as that of the lubricants, and they are best stored under the same conditions.

The laundering and disposal of used lubricants were described in detail in Chapter 6. The equipment required for laundering or collecting used oils will often be located in the lubricant stores area, so that suitable space and facilities should be incorporated in the planning. It should also be separated in some way from the new lubricant storage area, to avoid any possible confusion.

13.4 Dispensing and applying lubricants

The first and vital action before lubricant containers are opened is to clean the outer surfaces thoroughly to make sure that no dirt enters to contaminate the lubricant.

Oil drums are readily fitted with taps which can be screwed into the larger hole in the drum top in place of the threaded plug. The tap should preferably be the quick shut-off non-drip type which allows precise control of the amount of oil drawn off, even with very viscous oils. The tap is fitted when the drum is vertical, and the drum is then placed horizontally on a rack or stillage with the tap at the lowest position. The smaller upper plug is loosened to ensure that no oil leaks from it, and is then removed. The lubricant can then be drawn off by means of the tap.

If a barrel or drum is to be kept in the upright position, a hand or air-operated pump can be fitted to the larger threaded hole to enable the oil to be pumped out.

It is desirable to fit a piece of cloth or a loose wad of cotton waste over the ventilation hole to prevent dust or dirt from being drawn in, especially in a dusty atmosphere.

Grease containers present more difficulties because the whole of the top of a grease keg is a lid which must be removed before the grease can be dispensed. Every effort must be made to ensure that dirt or water does not enter while the lid is removed, and the lid must be replaced as soon as possible. Whereas oil can, to some extent, tolerate solid contaminants, which sink to the bottom of the drum, a grease will hold all the harmful solids which enter, and transfer them to the bearings.

Pumps are available for transferring grease to smaller containers or grease-guns, but a follower plate must be placed on top of the grease to press it down and ensure steady feed at the pump inlet.

The smaller dispensing containers used to transfer oil or grease from store to machine are generally quite well known. For oils there are the conventional oil-cans which are used to feed oil direct to a bearing or oiling point. They either have compressible walls which are pressed to expel oil from the spout, or they have a hand-operated piston or trigger which pumps a fixed volume of air into the can and so expels a similar volume of oil.

Larger quantities of oil are conveniently transferred by cans similar to those used for storing and dispensing paraffin. They have a hinged cover to keep out dirt, and either a fixed funnel-shaped spout or a screw-in spout on a flexible pipe. They come in sizes from 0.5 to 10 litres, the 10 litre size being similar in appearance to a garden watering can.

Where very large quantities of oil must regularly be supplied to certain points in the factory, it may be most efficient to install a pipeline from the store to the points of use.

Grease is usually transferred by means of grease guns, which are described in Chapter 7, and which can hold up to 2 kg of grease. Transfer of grease in intermediate pails is undesirable due to the difficulty of preventing contamination. If a supply of several kilograms or more of grease must be transferred from the lubricant store, for example to the reservoir of a centralized lubrication system, it may be best to buy it in suitable sized containers.

The exception to this general recommendation is where a fully enclosed pressurized or pumped transfer is used. The possibility of contamination is then largely eliminated. Suitable equipment includes portable power-operated pumps and portable grease-gun fillers, both of which can be filled directly from the keg or barrel in the lubricant store by means of a pump.

A portable power-operated pump consists of a grease reservoir, usually mounted on an integral trolley, fitted with one or more delivery nozzles and a suitable pump. For the softer greases up to No. 2, a hand or air operated piston-pump may be used. For stiffer greases a screw feed is suitable. Portable systems are available which will hold up to 25 kg of grease.

A portable grease-gun filler consists of a vessel like a modified grease pail fitted with a hand or air operated pump and a delivery nozzle. A follower plate is necessary for all but the most fluid greases.

Every item of dispensing equipment used for lubricants should be kept for one grade only, or thoroughly cleaned after each use. It is a waste of time, effort, and money to specify carefully the best oil for a machine, and then to allow that oil to become mixed with a less suitable one because of a shortage of oil-cans.

The importance of correct lubrication means that care must be taken over the selection and training of suitable lubrication staff, and the introduction of a proper system of lubrication control. It is a significant fact that in the former East Germany factories above a certain size were legally required to appoint a Lubrication Engineer, in the same way that many other countries require the appointment of a Safety Officer.

The importance of correct lubrication has been increasingly recognized in recent years, but there are still many plants where lubrication is carried out by either machine operators, machine minders, or labourers. There are problems with any of these situations. In general, modern machine operators are overwhelmingly concerned with maintaining output, and unless they have been trained to understand the need to use the right lubricant in the right quantities, they are likely to place more emphasis on rapid lubrication than on correct lubrication.

The unskilled worker will similarly need effective training and guidance in order to carry out the lubrication requirements satisfactorily. Understanding the need for correct lubrication requires some technical knowledge of the way bearings work. With proper training a responsible unskilled person should be able to carry out the lubrication task reliably, but not as one of a wide range of other duties.

In other words, whoever is required to carry out plant lubrication should be specially appointed and trained to do so. The number of staff required may vary from two in a small works or department to scores in a large integrated works. In all cases they should be able to specialize in lubrication.

The operation of a correct lubrication programme will be very much improved if a suitable system of lubricant stock control and lubrication instructions is introduced. One very useful technique is to use colour-coding, so that, for example, the colour allocated to a particular oil will be marked on the machine, the dispensing equipment, the central lubricant tank or drum, the rack position, the lubrication instructions, and the stock control cards. This system can become complicated in a large works where many different lubricants are used, but the lubrication staff should then be large enough to maintain effective control.

Major lubricant suppliers will carry out a lubrication survey of a plant, and design a complete system, from colour-coding and stores design to staff training. Very often a condition for carrying out such a survey is a commitment on the part of the user to purchase that supplier's products.

A user wanting to retain freedom to buy on the open market can arrange for a survey to be carried out by a specialist consultant, but will be charged a fee for the work. The cost may be recovered by being able to choose between competing lubricants.

It cannot be too strongly emphasized that the proper storage, handling, and application of lubricants will repay the same care and attention as would be given to any other high-quality precision component.

Chapter 14

Health and Safety

14.1 Overall safety of lubricants

It must be emphasized immediately that this chapter is not intended to give authoritative information about lubricant hazards or about safety precautions to be used with lubricants. For such information those responsible for lubrication matters, in conjunction with works Safety Officers and Medical Officers, should consult the detailed Safety Data Sheets provided by lubricant suppliers, and the information published by organizations such as Government Departments and the British Royal Society for the Prevention of Accidents (RoSPA).

The chapter is intended only to give a broad picture of the types of hazard which can arise in connection with lubricants, and the nature of the working practices which should be used.

Lubricants as a group are not particularly hazardous. Although most of them will burn, they are far harder to ignite than fuels and many solvents and chemicals. Although they can contribute to accidents, so can almost anything on Earth. Finally, although they can lead to health problems, they are far less dangerous to health than many materials which are used in much larger quantities.

Recent legislation in most developed countries has focussed public attention more closely on health and safety hazards, placed responsibility on suppliers, employers and employees to reduce risks, and resulted in more general availability of information about specific dangers and preventive actions.

Such legislation has done a great deal to improve work safety, but the greatest factor in avoiding accidents and industrial health problems is careful working and commonsense. This applies to lubrication as well as to most other activities of Man.

14.2 Flammability and explosions

One of the major problems in discussing fire-resistant materials is that different materials can be ignited readily under widely different conditions. For this reason it is difficult to classify lubricants simply as flammable or non-flammable. (The word 'inflammable' is not used in discussing fire-resistance because to people unfamiliar with English it could be dangerously misunderstood as meaning non-flammable. The words 'flammable' and 'inflammable' both mean that something can be made to burn. The word 'flammable' is preferable because it is not likely to be misunderstood.)

Combustion, or burning, is simply a form of rapid oxidation in which enough heat is produced to continue the oxidation. The burning material is broken down, usually to smaller molecules of highly oxidized combustion products.

For combustion to take place, there must be a flammable material, called a fuel, and an oxidant present. Oxygen gas is the most common oxidant, and it is of course present in air in high enough concentration to support combustion of many materials.

Combustion can be started by heat alone if the temperature is raised high enough to produce the necessary rapid oxidation. It can be started at a lower temperature if a flame is present, because the flame itself is at extremely high temperature and contains very reactive substances which encourage oxidation of the fuel.

Many test methods have been developed to assess fire resistance under different conditions, but there has been little agreement on standardizing any of them. It will probably be helpful to explain first some of the terms used in describing ignition, and then to describe the main types of test condition.

A 'flash' is an ignition which is very brief, lasting less than a second, and usually much less than a second. It occurs when a vapour is ignited but when no more vapour is immediately available to sustain the combustion, and the conditions do not enable anything else present to ignite.

A 'flame' is burning gas. A visible flame occurs when the burning is intense enough to cause incandescence, that is for bright light to be produced.

'Fire' is a self-sustaining combustion in which flames are produced. It differs from 'smouldering' which is self-sustaining combustion without flames. In both cases the heat produced by the combustion of some part of the material is sufficient to keep the temperature high enough to ignite a further part. The difference is that with fire the material burns to produce flammable gases and there is enough oxygen present to burn those gases. With smouldering the material burns but either produces no flammable gases or produces flammable gases but there is insufficient oxygen present to burn the gases. Fire or smouldering will not necessarily continue until all the flammable material is consumed. Sometimes the fire goes out, and the material may then be said to be self-extinguishing.

A 'spark' is a tiny fragment of material so hot that it is incandescent. Its temperature will probably be over 2000 °C.

An 'explosion' takes place when pressure builds up to such an extent that it cannot be contained or dispersed slowly. A container explodes when pressure builds up enough to rupture the container violently. An explosive is a material which builds up pressure so quickly that even when it is not in a vessel it has to disperse faster than the speed of sound, producing a shock wave.

The following are the important types of test which are often applied to lubricants.

(a) *Flash point*

Flash points are so often measured that they have already been described in Chapter 11. The flash point is the lowest temperature of a liquid at which the vapour above the liquid can be ignited by an open flame. In most cases it is the lowest temperature at which any ignition risk arises at all, and the risk may in fact be very slight. If a kerosine has a flash point of 50 °C, that means that it must be heated to over 50 °C, and a flame must be brought near it before its vapour will ignite.

The lightest mineral lubricating oils will usually have flash points over 120 °C, although some special types may be as low as 80 °C. The flash points of synthetic oils depend very much on their thermal stability. For example silicones and polyphenyl ethers, which are stable to quite high temperatures, have flash points over 270 °C. On the other hand, most other synthetic oils, even including phosphate esters and chlorinated diphenyls which are used for their fire resistance, have flash points around 170°–200 °C.

Flash points and other fire resistance data for different base oils are summarized in Table 14.1.

(b) *Spontaneous ignition*

If a substance is heated in air, a temperature level may be reached at which it ignites without any flame or spark having to be applied. This temperature is called its spontaneous ignition temperature, or SIT (also known as autogenous ignition temperature). If a similar test is carried out in pure oxygen the SIT is lower. This is normally only done where the substance is to be used in an oxygen system.

Table 14.1 Fire resistance data

Base oil	Flash point (°C)	SIT	Wick ignition	Spray ignition	Dieselling
Mineral	120–240	Low	Very flammable	Highly flammable	Very explosive
Di-ester	230	Low	Flammable	Very flammable	Explosive
Phosphate ester	200	Very high	Flammable	Difficult to ignite	Difficult to ignite
Silicone	300	High	Self-extinguishing	Difficult to ignite	Difficult to ignite
Polyglycol	180	Medium	Flammable	Flammable	–
Chlorinated diphenyl	180	Very high	Non-flammable	Non-flammable	Non-ignitable
Fluoro-ether	None	None	Non-flammable	None	Non-ignitable

Warning: In contact with *liquid* oxygen, even at very low temperatures, many materials which are not considered very flammable will ignite violently. This applies to almost all liquid lubricants, notable exceptions being the perfluorinated polyethers and fluorocarbons.

There is one peculiar effect in spontaneous ignition of liquids or gases which makes it difficult to obtain agreement on a standard test procedure. If a spontaneous ignition test of a liquid or gas is carried out in a larger vessel, the SIT will decrease. For this reason, tests have been carried out in 175 ml conical flasks, 250 ml conical flasks, 1 litre spheres, and 1 cu ft (i.e., 28 litre) spheres. The ASTM D–2155 test used a 250 ml flask, but it was recognized that a liquid in the open may ignite spontaneously at a lower temperature than in this test.

For this reason, the spontaneous ignition temperatures are listed in Table 14.1 only in relative terms. However, mineral oils and di-esters have consistently low spontaneous ignition temperatures, while phosphate esters, chlorinated and fluorinated silicones, chlorinated diphenyls, and fluoroethers consistently have very high spontaneous ignition temperatures.

Many other fire-resistance tests are in fact spontaneous ignition tests because no flame or spark is applied to cause ignition. They are designed to examine the effect of some special heating source, such as a hot manifold or a bath of molten metal, which is of particular importance to the user concerned.

Since test conditions are not widely standardized, it is not possible to make direct comparisons between tests, or between fluids tested by different methods. For example, petrol (gasoline) has a lower SIT than paraffin (kerosine) in a flask test, but it may be higher in a hot manifold test because at certain temperatures the petrol will evaporate too quickly to ignite.

(c) *Wick ignition*

A special type of hazard arises when a fibrous material such as fibre insulation, lagging, or cotton waste is damp with a flammable liquid. A thin film of the liquid forms on all the fibres, and as a result a small amount of liquid has an enormous surface area. A litre of oil might have a surface area of well over 40 square metres (440 square feet).

When heated, such as in the lagging of a steam pipe, this thin film may ignite. However, because there is little oxygen and little fuel present at any one point, the material does not burst into flame but continues to smoulder. Such a smouldering combustion may remain undetected for some time, but can act as an ignition source for other materials and lead to a more serious fire, as well as producing toxic or suffocating gases. Since it is the liquid which ignites, such problems often arise even in asbestos lagging.

To test for this effect, wick ignition tests are carried out. They usually consist of a system in which one end of a short wick is dipped into the test liquid under carefully controlled conditions, and a flame is applied to the other end.

Apart from the ease with which the liquid can be ignited, two other factors affect the results of a wick test. The first is the viscosity of the liquid, since a more viscous liquid will rise more slowly up the wick. The second is the nature of the combustion products formed from the liquid, since non-flammable solid combustion products can block the flow of liquid and prevent burning.

As a result, a phosphate ester which has a low viscosity will burn more easily in a wick than a silicone, which produces blocking deposits of silica when overheated.

The results of wick tests must therefore be interpreted with care and not applied to other combustion situations. They are useful for the particular case of lagging fires but also, in conjunction with other tests, help to give a broader overall picture of the fire resistance of different lubricants.

(d) *Spray ignition*

A liquid can be readily ignited when in the form of a fine spray because a large volume of air or oxygen can reach a small droplet of the liquid, and because there is no bulk of liquid in contact to cool the droplet. Since a spray is produced by liquid under pressure, there is often also a continuing supply of fuel available to keep the fire burning. Some catastrophic industrial and aircraft fires have been caused by ignition of a spray.

Tests for spray ignition are carried out by producing a spray of the liquid under pressure through a fine jet and igniting it by means of a flame or an electric arc. The result is stated in terms of the difficulty or ease of ignition, and the persistence of the flame. A high viscosity liquid is generally harder to ignite in a spray, simply because it is harder to produce a rapid supply of fine droplets.

Spray ignition tests tend to show that those liquids with low SITs ignite most readily, except for the complicating effect of viscosity.

The test is best used to assess fire risk for systems in which liquids will be present at high pressure, such as in a high-pressure hydraulic system. It also helps to give a broader picture of the fire resistance of different liquids.

(e) 'Dieselling'

Dieselling is a term used for ignition which takes place when a mixture of a vapour and an oxidant is compressed rapidly, in the same way as the fuel/air mixture in a diesel engine.

When a gas is compressed rapidly it becomes hot because of what is called adiabatic compression. (Adiabatic means without loss of heat to or from the outside.) It is this heating which causes the mixture in a diesel engine to ignite, and the same situation can arise in some other circumstances. It is not a common phenomenon, but it has been known to cause serious accidents in aircraft undercarriage oleos and in the crankcases of reciprocating compressors if the breather becomes blocked.

Testing for dieselling risk is carried out in a piston-cylinder arrangement, usually with a facility for controlling and increasing the outside temperature. The liquids most resistant to dieselling are those which have high flash points and high spontaneous ignition temperatures, such as silicones and fluoroethers. These are, unfortunately, not among the best lubricants for a piston-cylinder arrangement. Where possible, therefore, it may be effective to use a less explosion-resistant lubricant and to fill the chamber with nitrogen instead of air to prevent dieselling.

14.3 Accidents involving lubricants

There is a relatively high risk of mechanical accidents occurring where loads are carried by moving surfaces. These are exactly the situations where lubricants are used, so that it is natural that lubricants are often involved where accidents occur in machinery. In addition, most common lubricants can burn, so that the risk of fire is added to the dangers of accidents in lubricated systems.

The use of the wrong lubricant, or faults in lubricant supply, may contribute directly to premature failures of lubricated systems such as gears and bearings. For example, in a steam turbine in a chemical plant there was a failure in the lubricating oil pump, which resulted in an interruption in the oil flow. The standby pump took over almost immediately, but the temporary loss of oil wrecked a tilting pad thrust bearing. This allowed the shaft to move axially, and the turbine wheel was also wrecked and a leak of high-pressure steam occurred. The plant was then shut down without further damage.

Often, however, the lubricant does not contribute in any way to the failure, but only to the consequences. For example, the propellor shaft of a small ship suffered a fatigue failure just forward of the stern tube bearing. This caused overloading and misalignment at the forward bearing, which seized. The teeth were stripped from the splined drive at the gearbox, and before the engine could be shut down the damaged spline overheated and ignited the oil leaking from the bearing, causing a fire.

The temperatures reached by a damaged bearing or coupling can be very high. In one damaged thrust bearing, the temperature was shown to have reached over 1100 °C. With lubricant present to act as a fuel, these high temperatures can very easily cause serious fires, especially where the plant is handling some highly flammable gas or liquid. One such fire alone resulted in damage estimated at over £3 000 000.

Oil is more likely than grease to lead to serious fires, but seizure of grease-lubricated conveyor bearings in coal mines has been blamed for several potentially dangerous fires. A similar problem is believed to have contributed to the disastrous fire in the Kings Cross Underground Station in London.

There is an obvious second way in which lubricants can cause accidents. Lubricants are, by their nature, slippery, and when they occur on floors, ladders, roads or ships' decks, they can lead to unpleasant and even serious accidents.

For this reason, special care must be used in handling lubricants to avoid spillages and to clean them up rapidly when they occur. This is particularly important in lubricant stores or cellars, or any other place where large quantities of lubricants are used. Small spills which are not cleared up lead both to a sloppy work attitude and to an increased risk of further accidental spills.

Such areas should have floor grills wherever there is likely to be continuing spillage, such as beneath oil tank taps. In time metal grills can become highly polished, and even a thin film of oil on a polished grill can cause a hazard. Grills should therefore be cleaned with non-flammable solvent or chemical cleaners, and roughened or replaced when necessary.

On other floor areas oil spills should immediately be covered with absorbent powder or granules. These generally give high friction and in this way reduce the risk of slipping or falling until the granules can be swept up and the floor properly cleaned.

14.4 Health aspects of lubricants

This brief description of some of the health aspects of lubricants is not intended to be either comprehensive or authoritative. Authoritative data on health hazards for all sorts of chemical materials are published by most governments and other public bodies, and hazards of particular lubricants are described in the relevant manufacturers' Safety Data Sheets. Further general information is contained in some of the books listed in the Bibliography on p. 272.

On the whole, lubricants are rather stable, inert materials, and the health problems which arise from their use tend to be long-term rather than immediate. Acute (i.e., rapid or immediate) toxicity is more likely to arise from the more reactive additives than from the base oils, especially where additives are present in high concentrations, such as in soluble oils and some diesel engine lubricating oils.

Some of the longer term health problems include cancer and dermatitis. There is strong evidence that these occur only when the victim has been careless and has allowed his skin to be in contact with oil or oil-soaked clothing for very long periods. Nevertheless, the lubricant manufacturers have made great efforts to identify and remove the dangerous components. It is hoped that the risk of cancer has now been completely eliminated. However, because it can take thirty years for certain cancers to develop it will be many years before the danger will be known to have disappeared.

It should be emphasized immediately that the health risks associated with lubricants can probably be avoided completely by care in avoiding prolonged contact with the skin. Oil on the hands should be cleaned off immediately, and oil-wet clothing changed.

Table 14.2 shows some of the types of health hazard which can arise with specific lubricants. These will be described in more detail in the next few pages, but for authoritative information the manufacturers' product Safety Data Sheets should be consulted.

(a) *Mineral oils*

Acute toxicity of mineral oils is very low, and even with the most dangerous of mineral oils it would probably be necessary to drink over

Table 14.2 Health hazards of lubricant materials

Oil type	Acute toxicity	Dermatitis	Cancer
New mineral oil	Very slight	Care required	Believed to be disappearing
Used mineral oil	Some risk	Care required	Care required
Synthetic hydrocarbon	Very slight if any	Care required	None reported
Di-ester and polyol ester	Slight	Care required	None reported
Phosphate ester	Some risk	Care required	None reported
Silicone	Non-toxic	Little risk	None reported
Polyglycol	Believed to be low	Believed to be low	None reported
Chlorinated diphenyl	Irritant vapour when hot	Care required	None reported
Fluoroether	Toxic vapour when over-heated	Not known	None reported
Soluble oil	Care required	Care required	As for mineral oil
Grease	Very slight	Care required	Probably very little risk
Additives	Some risk	Little risk	Little if any risk

half a litre before there would be severe risk. There would probably be severe vomiting if anything approaching such a quantity was drunk. Absorption through the skin or by inhaling oil mist is also fairly innocuous. An American recommendation limits the concentration of soluble oil mist in the air to less than 5 mg per cubic metre of air. This is designed for comfort rather than avoiding toxic effects, and concentrations many times higher for 8 hrs per day for years have been shown to have no adverse effect.

The major health problem with mineral oils is certainly dermatitis. This is an unpleasant inflammation of the skin which can be hard to get rid of. Mineral oils will slightly dissolve the natural fats in the skin, and if any part of the skin is wet for a long period with mineral oil it will tend to dry and even crack. Oil can also block the pores in the skin, preventing perspiration, and thus blocking up harmful body waste products. Any of these factors can lead to dermatitis, and the best prevention is cleanliness. Older people and fair-skinned people seem to be more likely to develop dermatitis.

A more serious result of mineral oil contamination is skin cancer, but fortunately this is also very rare. The main cause is the presence of what are called Poly-Cyclic Aromatic Hydrocarbons (PCAHs), which are known to be carcinogenic. Solvent extraction of mineral oils removes

PCAH, and such refining techniques are becoming universally used for lubricant base oils.

Scrotal cancer has been shown to be caused by very long contact with mineral oils. The cause is again almost certainly PCAH, and improved refining techniques should get rid of the problem. It can certainly be avoided completely by care in avoiding prolonged contact with oil.

When hands become wet with oil or grease they should be cleaned as soon as possible using warm water and soap, mild detergents or proprietary skin-cleaning products. No solvents, white spirit, paraffin or petrol should be used, as they will also tend to remove skin fats. Abrasive household cleaners will also damage the skin. Where frequent wetting with oil is likely, barrier creams may be helpful, mainly because they make it easier to remove the oil by mild methods, but they are not as effective as gloves, and can cause problems. They should be chosen to match the sort of oil involved (e.g., neat oils or soluble oils), and proper advice should be obtained on their use.

If the hands become dry because of frequent contamination and cleaning, a good hand cream should be used. This will help to avoid skin cracking, and more serious infections. Men are often reluctant to use such creams because most proprietary grades are perfumed, but plain creams can be obtained.

If clothing becomes wet with oil it should be changed as quickly as possible. Oil-wet clothing is probably the main cause of skin cancer and particularly scrotal cancer, as it keeps oil in contact with the skin for long periods, and also rubs the skin causing sensitivity and inflammation. For workers in lubricant stores aprons are recommended which have an impermeable under layer and an absorbent outer layer.

A final preventive procedure is to report any skin complaints immediately they occur, and to cover any cuts or scratches with oil-resistant finger-stalls or plasters. Any skin damage gives an open door for oil and infections to enter, and such damage should be cleared up as quickly as possible, if necessary keeping the victims away from lubricating oils.

(b) *Synthetic oils*

In general, synthetic oils have low toxicity and no reported carcinogenic effects. Their mild toxic effects will, however, differ from one class to another. This again shows the desirability of manufacturers differentiating between them, and not simply labelling products as 'synthetic'.

Synthetic hydrocarbons include several different chemical types, but none are reported to have any acute toxic effects or carcinogenicity. They are in fact similar in some respects to medicinal paraffin, except that there is some risk of dermatitis because of a tendency to dry and crack the skin. Care should therefore be taken to avoid prolonged contact. However, one class which is used in small quantities in arctic engine oils and refrigerator oils is that of the alkylbenzenes. These are likely to have some toxicity associated with the presence of the aromatic (benzene) group.

Diesters and polyol esters have slight toxicity and no reported carcinogenicity. They can cause dermatitis, and care should be taken to avoid prolonged contact.

Polyglycols are believed to have low toxicity and low risk of dermatitis, but they are irritant on contact with the eyes.

Phosphate esters are toxic and there is a risk of dermatitis, so care should be taken to avoid ingestion or prolonged skin contact.

Fluoroethers and chlorinated diphenyls both produce toxic gaseous products when overheated to decomposition, and dermatitis can arise from prolonged contact with chlorinated diphenyls.

Silicones are for practical purposes completely harmless.

(c) *Additives*

Additives are more or less reactive chemicals, whereas most base oils are chemically very inert. Additives therefore tend to be more irritant and more toxic than base oils. Additives are usually used in small concentrations. Consequently, any risk is much reduced, except in lubricants which contain high concentrations of additives, such as soluble oils and engine oils. Nevertheless, all additives are now carefully screened for toxicity, and those with higher toxicity, such as lead naphthenates, chlorinated naphthalenes, and orthophosphates, have been largely replaced.

Phosphates are used as EP and anti-wear additives. They can occur in three forms, called ortho, meta, and para-phosphates. Until about 1965 no particular distinction was made between the three forms. It then became apparent that the ortho form was highly toxic in large amounts, causing damage to the nervous system, resulting in blindness or paralysis. Such serious damage was not found with normal lubricant use, but cases occurred where unscrupulous merchants in undeveloped regions blended hydraulic fluid containing tricresyl phosphate (TCP) into alcoholic drinks, and large numbers of cases of blindness resulted. The concentration of the ortho form has since been restricted to less than 1 percent of the TCP. Since the normal concentration of TCP in a lubricant or hydraulic fluid is less than 1 percent, the concentration of the ortho form is less than 0.01 percent of the oil. This is harmless in any quantity of oil which could reasonably be swallowed.

Lead naphthenate soaps have been used for many years as EP

additives in industrial gear oils. It was found that slight increases in lead concentrations occurred in some steel industry workers who handled large quantities of those types of gear oil. Although no harmful results were ever demonstrated, there has been a general move away from lead naphthenate additives to sulphur-phosphorus types.

Additives are likely to be more of a hazard to lubricant manufacturers than to users, because in manufacture large quantities of neat additives will be used. For the user the usual precautions against too much wetting by oil and against accidental swallowing will avoid any risk from additives. If for any specialized application it is necessary to use the more toxic additives, modern regulations and practice would ensure that clear warnings are given on the package and in Safety Data Sheets, and safety precautions clearly indicated.

(d) *Soluble oils*

There are four special hazards in the use of soluble cutting oils, but none of them is serious if commonsense is used.

Neat soluble oils have high concentrations of additives, but these are very much reduced once the oil has been dispersed in water. Toxicity problems should, therefore, be low as far as the machine operator is concerned.

Soluble oils are particularly prone to microbiological attack, i.e., growth of bacteria and microscopic fungi, because they contain large amounts of water. These do not in themselves cause a hazard to human beings because the organisms present are almost entirely harmless to man. The problem arises from the biocides which are added to prevent microbiological attack. Biocides are materials which have strong toxicity to micro-organisms, and they also have some toxicity to Man. The usual precautions against ingesting the oils will prevent trouble.

Other water-containing oils such as invert emulsions are also sometimes treated with biocides, and the same precautions should be taken.

High-speed cutting processes can lead to the formation of a fine mist of cutting oil, especially where the tool or workpiece becomes hot. Care should be taken not to breathe in this mist. A good guide is that if a coolant mist is visible the operator should not hold his face near it or should use a mask. Conversely, if no mist is visible the air is safe to breathe even when the operator is working full-time at the machine.

Soluble oils contain high concentrations of detergents to enable the oil to form a stable dispersion in water. These detergents, like soap and household detergents, can extract fat from the skin and cause drying and cracking, thus encouraging dermatitis. Care should be taken to avoid long contact with soluble oils, in the same way as described earlier for mineral oils.

One final problem in machining is that fine swarf in the air, in the coolant, or in cleaning rags, can cause fine cuts in the skin and encourage infections and dermatitis. Contaminated hands should preferably be washed rather than wiped clean, using a good-quality industrial hand cleaner. This should be supplied by means of a dispenser, not in open containers, to avoid contaminants entering the cleaner.

(e) *Used oils*

There is evidence that used oils, especially from hot applications such as engines and machining contain increased quantities of carcinogenic PCAHs. They also contain aldehydes among the oxidation products and these can be toxic, although their concentrations are very low. Used gasoline engine oils also contain PCAHs from combustion products, and lead from octane improvers, while used machining oils contain fine swarf particles. There is, therefore, some increase in risk with used oils, although there is no evidence that any harmful effects have occurred.

The increasing re-use of used lubricants causes concern about a possible build-up of PCAHs. Where simple laundering is carried out with industrial hydraulic fluids or gear oils, no problem will arise because such oils do not become very hot in use, and do not have increased PCAH content.

Laundering and topping-up of soluble oils with neat oil will lead to a build-up of some additives, and this is likely to be more of a problem than any increase in PCAHs. Re-use in this way is more common with large automatic machine-tools and integrated production lines, where contact of the operator with cutting oils is very much reduced. In large jobbing shops, where operators are in closer proximity to the machining operations, some guidance or restriction might eventually be desirable on the extent of re-use of cutting oils. So far the extent of re-use has been fairly small, and there has been no indication of harmful effects.

The main processes used for re-refining of used lubricants will, to some extent, remove PCAHs. The solvent processes such as the liquid propane process will almost completely remove them, and the re-refined oil will be as safe as modern solvent-refined oils fresh from the refinery. The acid-clay process results in some removal of PCAHs, but not to the low levels of solvent-refined oils.

Most re-refined oils are used as engine lubricants or blended into fuel oils, so that the degree of human contact is small. The usual precautions against prolonged contact and accidental swallowing should be taken, and should be completely effective.

(f) *Greases*

Greases are used in smaller quantities than oils and are not easily spilled, so the health risks are much lower than with oils. They will stick more firmly to the skin, and many of them contain soaps, so that frequent contact with greases can cause dermatitis. The risk will arise mainly with those who regularly handle greases, either in filling grease-guns or grease reservoirs, or in packing bearings.

For filling grease-guns or reservoirs, spatulas or paddles should be used to handle the grease, and industrial rubber gloves can be worn to protect the hands. For packing bearings it is common practice to use the fingers, and the best way to avoid contact is to use the lightweight or 'surgical' rubber gloves used in chemical laboratories or in instrument assembly.

In conclusion, to keep the health hazards associated with lubricants in perspective, serious health hazards from lubricant handling are far less common than serious injuries from industrial accidents, and can be avoided completely with reasonable care and commonsense.

Bibliography

Allinson, J. P., 1973, *Criteria for quality of petroleum products*, Applied Science Publishers, London, UK.

Annual Book of ASTM standards: Parts 23–25, *Petroleum Products and Lubricants*, American Society for Testing and Materials, Philadelphia, USA (published annually).

Austin, R. M., Nau, B. S., Guy, N., and Reddy, D., 1974, *The seal users handbook*, BHRA Fluid Engineering, UK.

Barwell, F. T., 1956, *Lubrication of bearings*, Butterworths, London, UK.

Boner, C. J., 1976, *Modern lubricating greases*, Scientific Publications (GB) Ltd, Broseley, Shropshire, UK.

Braithwaite, E. R. (editor), 1967, *Lubrication and lubricants*, Elsevier, London, UK.

Clauss, F. J., 1972, *Solid lubricants and self-lubricating solids*, Academic Press, New York, USA.

Gunderson, R. C. and Hart, A. W., 1962, *Synthetic lubricants*, Reinbold, New York, USA.

Ku, P. M. (editor), 1973, *Interdisciplinary approach to liquid lubricant technology*, NASA SP–318, National Aeronautics and Space Administration, Washington, DC, USA.

Lansdown, A. R., 1994, *High temperature lubrication*, Mechanical Engineering Publications Limited, London, UK.

Lansdown, A. R., 1973, *Molybdenum disulphide lubrication*, ESRO CR-402, European Space Research Organisation.

Mortier, R. M. and Orssulik, S. T. (editors), 1992, *Chemistry and technology of lubricants*, Blackie and Son, Ltd, Glasgow and London, UK.

Neale, M. J. (editor), 1992, *The tribology handbook, Second Edition*, Butterworth Heinemann, London, UK.

Nica, A., 1969, *Theory and practice of lubrication systems*, Scientific Publications, Broseley, Shropshire, UK.

O'Connor, J. J., Boyd, J., and Avallone, E. A., 1968, *Standard handbook of lubrication engineering*, McGraw-Hill, New York, USA.

Standard methods for analysis and testing of petroleum and related products, John Wiley and Sons (on behalf of the Institute of Petroleum, London), (Published annually).

Summers-Smith, J. D., 1994, *An introductory guide to industrial tribology*, Mechanical Engineering Publications Limited, London.

Index